An OBE Approach to Logic Circuits and Digital Design: Step by Step

An OBE Approach to Logic Circuits and Digital Design:
Step by Step

Ma. Beatriz Lacsamana

MBLacsamana
2015

Copyright © 2015 by Ma.Beatriz Lacsamana

All rights reserved. This book or any portion thereof may not be reproduced or used in any manner whatsoever without the express written permission of the publisher except for the use of brief quotations in a book review or scholarly journal.

First Printing: 2015

ISBN 978-1-312-82832-2

Dedication

FOR YOU

TABLE OF CONTENTS

Ma. Beatriz Lacsamana iii

TABLE OF CONTENTS vii

Acknowledgements ix

Introduction 1

CHAPTER 1 : HISTORY OF DIGITAL ELECTRONICS 2
    TIMELINE 3
        B: Second Generation: Transistors 4
        C. Third Generation: Integrated Circuits 5
        CHAPTER 1 END OF THE CHAPTER ACTIVITIES 7

CHAPTER 2 BUILDING BLOCKS OF COMBINATIONAL LOGIC CIRCUITS 8
    I – BASIC GATES 8
    II - THE UNIVERSAL GATES NAND AND NOR 11
    III - LEVELS OF INTEGRATION OF INTEGRATED CIRCUITS 14
        CHAPTER 2 END OF THE CHAPTER ACTIVITIES: 15
    EXPERIMENT # 1: BASIC LOGIC GATES 17

CHAPTER 3 - REVIEW OF BOOLEAN ALGEBRA 23
    PRINCIPLE OF DUALITY 24
    COMPLEMENT OF A FUNCTION 25
    SIMPLIFICATION OF BOOLEAN EXPRESSIONS 26
        CHAPTER 3 END OF THE CHAPTER ACTIVITIES: 30
    EXPERIMENT 2 : SIMPLIFICATION OF A FUNCTION USING BOOLEAN ALGEBRA 31

CHAPTER 4 : MULTILEVEL COMBINATIONAL LOGIC CIRCUITS 35
        CHAPTER 4 END OF THE CHAPTER ACTIVITIES 42
    EXPERIMENT 3 : UNIVERSAL GATES 45

CHAPTER 5 - STANDARD FORMS and TWO-LEVEL IMPLEMENTATION 48
    SUM OF PRODUCTS (SOP) OR SUM OF MINTERMS (SOM) 49
    PRODUCT OF SUMS (POS) or PRODUCT OF MAXTERMS (POM) 52
    PROCEDURE FOR CONVERSION BETWEEN FORMS: 54
    Two-Level Implementations of Standard Forms 56
        Sum of Products Expression (SOP): 56
        Product of Sums Expression (POS); 57
        CHAPTER 5 END OF THE CHAPTER ACTIVITIES : 60
    EXPERIMENT 4 - STANDARD SOP TO POS: CONVERSION BETWEEN FORMS 61

CHAPTER 6 – MINIMIZATION USING KARNAUGH MAPS 64
    2- VARIABLE K-MAPS 64
    3-VARIABLE K-MAPS 67
    4-VARIABLES K-MAP 71
        CHAPTER 6 END OF THE CHAPTER ACTIVITIES : 79
    EXPERIMENT 5 : MINIMIZATION USING KARNAUGH MAPS 80

CHAPTER 7 - COMBINATIONAL LOGIC CIRCUIT ANALYSIS 83
CHAPTER 7 END OF THE CHAPTER ACTIVITIES: 91
EXPERIMENT 6 – ANALYSIS OF A COMBINATIONAL LOGIC CIRCUIT (CLC) - FULL SUBTRACTOR CLC ANALYSIS 94

CHAPTER 8 - COMBINATIONAL LOGIC CIRCUIT DESIGN 96
DESIGN PROCEDURE : STEP BY STEP 96
CODE CONVERTERS 97
ADDERS 103
SUBTRACTORS 106
DECODERS 109
ENCODERS 114
MULTIPLEXERS 116
DEMULTIPLEXER OR DEMUX 123
CHAPTER 8 END OF THE CHAPTER ACTIVITIES: 124
EXPERIMENT 7 : CLC DESIGN – CODE CONVERTER 126
BCD TO SEVEN SEGMENT DECODER 126
EXPERIMENT 8 : CLC DESIGN – ADDERS 130
FULL ADDER 130
EXPERIMENT 9 : CLC DESIGN – DECODERS : FULL ADDER IMPLEMENTED USING DECODERS 133
EXPERIMENT 10 : CLC DESIGN –MULTIPLEXERS 136
FULL ADDER IMPLEMENTED USING MUX 136

CHAPTER 9 : BUILDING BLOCKS OF SEQUENTIAL LOGIC CIRCUIT 139
9.1 FLIP-FLOPS 140
9.1.1 THE S-R (Set-Reset) Flip Flop 140
9.1.2 THE JK FLIP FLOP 143
9.1.3 THE T-FLIP FLOP 146
9.1.4 The D Flip Flop 146
Experiment # 11 : BASIC FLIP FLOP OPERATION 150

CHAPTER 10 : THE ANALYSIS OF OF SEQUENTIAL LOGIC CIRCUITS 153
ANALYSIS PROCEDURE: STEP BY STEP 154

CHAPTER 10 END OF THE CHAPTER ACTIVITIES: 163
EXPERIMENT 12 : THE ANALYSIS OF A D FLIP FLOP CIRCUIT 167
EXPERIMENT 13 : THE ANALYSIS OF A T FLIP FLOP CIRCUIT 170
EXPERIMENT 14 : THE ANALYSIS OF A JK FLIP FLOP CIRCUIT 173

CHAPTER 11 : THE DESIGN OF SEQUENTIAL LOGIC CIRCUITS 176
STEP BY STEP PROCEDURE : 176
FLIP FLOP EXCITATION TABLES 177
CHAPTER 11 END OF THE CHAPTER ACTIVITIES: 191

CHAPTER 12 - NEXT STEPS : INNOVATE, INNOVATE,INNOVATE! 192

Notes 195

References 197

Acknowledgements

I would like to thank everyone without whose help this book would never have been completed.

Thank you for the opportunity, your kindness, patience and guidance.

-

Introduction

Many years ago as a young student of Electronics and Communications Engineering, books for our technical subjects were difficult to find. Today, one can look up practically everything online. However, the myriad of information have made many students more confused than ever. The question that arise then is how do you guide them so that they just hit the goal and minimize the confusion caused by the overwhelming bits and bytes of information? The key is to adopt a framework much like Outcomes Based Education (OBE) so that the students keep their focus on the goal or outcome. OBE is very much like cooking, you think about the dish and prepare everything that you need to get that dish on the plate. Such is the inspiration behind this book which is designed with the end in mind. The goal is for the students to apply the skills learned to analyze and design digital circuits for relevant real-life applications. The last chapter of the book is a challenge for the students to complete with their very own mini-digital circuit project. This should then catalyze them to design and build more complex digital circuits.

Chapter 1 starts with a short introduction of how it all started highlighting and remembering the early fathers of the field. This is at the same time instilling the value of respecting the legacy of those who came before us. It also teaches the lesson that everything is fleeting especially in the field of technology . The idea is to guide the students to make the right choices and to learn from the highs and lows of history. The content provided is succinct but redirects the students to pertinent online sites in activities called "LOOK UP ACTIVITIES". Opportunity to meet the students where they are (which is in their gadgets)is also provided by encouraging them to make little apps related to the chapter. This activity we coined "STEP UP FOR YOUR APPS". To further sustain the students' interest and to reward them for a job well done another activity called "PICTURE, PICTURE" is included where the students documents several poses with their working model . It is amazing how interested this generation is in pictures such that we now have new words like "selfie" and "groupie".

The flow of the succeeding chapters start with the building blocks, analysis and then design. This primer on Digital Design categorizes the logic circuits in the book as Combinational Logic Circuits (CLC) and Sequential Logic Circuits (SLC).The tools interspersed as necessary. Experiments and Simulations at the end of the chapter are included for the students to verify their theoretical circuits on the breadboard and/or simulators. I observed the students during my many years of teaching were excited and inspired when they see the equations on their notes come to life in their breadboards and simulators I have employed this technique in the presentation of the equations and experiments that we tested in our classes. The chapters are filled with Illustrative Problems to explain clearly the concepts. The opportunity to LOOK UP is maximized even in the experiments where the students are encouraged to hunt for data sheets and other pertinent information online. This is meeting the students where they are at the same time keeping their focus by detailing the learning outcomes of each chapter and experiment.

It is my hope that the readers of this book will find this material both useful and enjoyable.

CHAPTER 1 : HISTORY OF DIGITAL ELECTRONICS

Learning Outcomes:

1. To appreciate the evolution of digital electronics and the importance of history.

2. Share insights on the key players of each generation and highlight the notable traits of the innovator and circumstances that led to the important discovery of the era.

3. To appreciate the legacy of the early fathers of the field.

4. To recognize the tipping points that drove the next wave.

5. To identify contemporary devices, gadgets and technologies.

6. To illustrate understanding by translating relevant information into a simple interactive app.

KEYWORDS

Digital Electronics is the implementation of two-valued logic using electronic logic gates such as gates, or gates and flip flops. [1]

Logic gate is a device, usually an electrical that performs one or more logical operations on one or more input signals. Logic gates are the building blocks of digital technology. [2]

Flip Flop is an electronic circuit having two stable conditions, each one corresponding to one of two alternative input signals [3]

Semiconductor is a substance, as silicon or germanium, with electrical conductivity intermediate between that of an insulator and conductor: a basic component of various kinds of electronic circuit element (semiconductor device) used in communications, control, and detection technology and in computers.[4]

Silicon is the material that is used as a base or substrate for most Integrated Circuits. [5]

Transistor is an electronic device that controls the flow of an electric current, most often used as an amplifier or switch. From transfer + resistor, so called because it transfers an electrical current across a resistor. Coined by John Robinson Pierce Of Bell Labs. It has replaced the Vacuum Tubes in most circuits since it is much smaller, more robust, and works at a much lower voltage. Transistors and other components are interconnected to make complex integrated circuits such as logic gates, microprocessors and memory [6]

Integrated Circuit is a device made of interconnected electronic components, such as transistors and resistors that are etched or imprinted onto a tiny slice of a semiconductor material such as silicon or germanium. An integrated circuit smaller than a fingernail can hold millions of circuits. Also called chip, or microchip. [7]

Tipping Point is the point at which an issue, idea, product, etc., crosses a certain threshold and gains significant momentum, triggered by some minor factor or change. It is the culmination of a build-up of small changes that effects a big change. [8]

TIMELINE

The history of digital electronics evolved in three generations. It started with the invention of the Vacuum Tube (VT) Diode acting like a switch. It can implement on and off operations . However the VT has many disadvantages which includes its large size, power inefficiency and high operating temperature. The smaller transistor solved these inefficiencies. It was the second generation. The need to miniaturize and to consolidate the devices together into one space led to the development of Integrated Circuits, the third generation.

A. First Generation: The Vacuum Tube (VT) Diode

In 1904, John Ambrose Fleming invented the VT Diode. It is a rectifier which passes current only in one direction. When it passes current in one direction it is ON .It is OFF when it prevents current to pass in the opposite direction. This is an early example of an electronic switch.

A.1 John Ambrose Fleming (1849-1945:) Father of Modern Electronics, Inventor, Educator, Pioneer, Innovator and Author

John Ambrose Fleming is first and foremost an inventor. He invented the VT Diode which laid the basis for modern electronics. He was under the tutelage of Maxwell at St John's College in Cambridge where he received his Doctor of Science Degree. Concurrent to his consulting work , he was a professor and chair of the Electrical Engineering at University College, London (UCL). He also did *pioneering work* in the field of education and was *responsible for new teaching methods* such as incorporating experimentation and laboratory work in the classroom setting. His contribution was material in the first transatlantic radio transmission in 1901. His focus on the work at hand was remarkable . The tipping point was when Fleming finally found a particular use for the Edison Effect which he used as the basis for his Fleming Valve or the VT Diode. The ramifications of the Fleming Valve were myriad and far-reaching. It was a key component of radios for nearly three decades, until it was replaced by the transistor, and was integral to the development of television, telephones, and even early computers. Fleming established the basis for electronics. As Orrin E. Dunlap, Jr., quoted Fleming as modestly commenting in Radio's One Hundred Men of Science ***The little* things of today may develop into great things of tomorrow.**.[9]

A.2 List of Vacuum Tube Computers

Table 1 First Generation -Vacuum Tube Computers [10]

Date	Computer Name	Remarks
1942	Atanasoff - Berry Computer (ABC)	
1943	Colossus	First programmable computers.
1946	ENIAC	18,000 VT, 200 KW, over 30 Tons
1948	Manchester Small-Scale Experimental Machine	First stored program computer.
1949	Manchester Mark 1	First index registers.
1949	BINAC	First stored program computer to be sold
1950	Pilot Ace	Based on a full-scale design by Alan Turing
1951	LEO I	First computer for commercial applications.
1958	EDSAC 2	First computer to have a microprogrammed control unit and a bit slice hardware architecture.
1958	AN/F-SQ-7	Largest Vacuum Tube Computer ever built for Project SAGE
1959	Rice Institute Computer	Operational, 1959-1971, 54 bit tagged architecture
1962	BRLESC	1727 tubes and 853 transistors

The Vacuum tube, however has many limitations to include the requirement of a heater for each cathode element, bulkier and heavier, higher operating temperatures and generally inefficient.

B: Second Generation: Transistors

To improve on the inefficiencies of the Vacuum Tube, Bell Labs commissioned a team of the best and the brightest engineers and physicists. The team leader of the solid state physics group was William Shockley, under him was Bardeen and Brattain. According to sources Shockley worked independently on his idea of field effect transistor whereas Bardeen and Brattain proceeded on their own model which is the point-contact transistor. The two came up with the first working model which made Shockley both happy and furious. To protect his position as leader of the team, he proceeded out of anger and creativity and came up with a better device which is more practical and can be easily manufactured. However, the three shared the Nobel Prize for Physics in 1956 for the invention of the transistor. Shockley founded Shockley Semiconductors. Later, some of his people would leave his company to put up Fairchild Semiconductors and eventually Intel Corporation. These companies were the pioneers of Silicon Valley, so called because of the material used in the manufacture of transistors and eventually integrated circuits. The transistor, in addition to its main function as an amplifier can also be operated as a switch, ON in the saturation region and OFF in the cut-off region.

B.2 Key Advantages of Transistors over Vacuum Tubes [11]

1. No power consumption by a cathode heater;

2. Small size and minimal weight, allowing the development of miniaturized electronic devices.

3. Low operating voltages

4. No warm-up period for cathode heaters required after power application.

5. .Lower power dissipation and generally greater energy efficiency.

6. Higher reliability and greater physical ruggedness.

7 .Extremely long life. Some transistorized devices have been in service for more than 50 years.

C. Third Generation: Integrated Circuits

According to nobelprize.org , *"there was a problem of numbers called the tyranny of numbers. Advanced circuits contained so many components and connections that they were virtually impossible to build. Jack Kilby of Texas Instruments (TI) found a solution to this problem by building smaller circuits. As a new employee at TI, Kilby had no vacation like the rest of the staff. Working alone in the lab, he saw an opportunity to find a solution of his own to the miniaturization problem. Kilby's idea was to make all the components and the chip out of the same block (monolith) of semiconductor material. For which he received the Nobel Prize in Physics in the year 2000. He also led the team that invented the hand-held calculator."* [12]

C.1 JACK KILBY

Jack Kilby is a multifaceted electrical engineer who started another breakthrough in miniaturization through the invention of the integrated circuit. His creativity and talent is widely demonstrated in the many inventions that he had throughout his career. To his credit, the world owe the following inventions:[13]

Plug-in Circuit Units, Semiconductor Structure Fabrication, Miniature Semiconductor Integrated Circuit, Miniature Semiconductor Network Diode and Gate, Miniaturized Electronic Circuits, Miniaturized Self-contained Circuit Modules, Semiconductor Structure Fabrication, Thermal Printer and Miniature Electronic Calculator.

C.2 ROBERT NOYCE

Robert Noyce shared the Nobel Prize for Physics with Jack Kilby for the invention of the Integrated Circuit. He co-founded Fairchild Semiconductors and Intel Corporation. If Jack Kilby is a purely technical person, Robert is both a techie, a businessman and a politician. He is noted for his business and political prowess. He was nicknamed the "mayor of Silicon Valley"

C.3 Advantages of the Integrated Circuit

1. It is small in size, occupies less space and allows for the construction of smaller devices and gadgets.

2. It consumes less power, therefore more efficient.

3. It is cheaper.

4. It is faster.

5. It requires lesser external wiring connections.

An interesting timeline which vividly describes the evolution of the Integrated Circuit can be found at chiphistory.org.

Gordon Moore, co-founder of Intel observes that the density of transistors on a per square inch of Integrated Circuit doubles every two years at the same time the price being halved. This is called Moore's Law.

CHAPTER 1 END OF THE CHAPTER ACTIVITIES

1. Form small groups and share among yourselves your insights about the early fathers of electronics. Highlight their traits and characteristics. Who is the figure that you admire most? Why?

2. Cite examples from your personal observation or from web sources proofs of Moore's Law.

3. What are some modern day gadgets which are made possible by digital electronics?

4. What are wearables? What are smart cars? What is IoE? What do you think is the next big thing?

LOOK UP :

1. Look for the pictures and life story and achievements of the early fathers.

2. Look for the other key players in the evolution of digital electronics. How important is their contribution?

3. Look for timelines, pictures, history, movies about the history of digital electronics. Highlight the one that strike you the most and and tell your team why.

REFLECTION PAPER :

1. Assuming that you are given the chance to live the life of the early fathers, who among them will you be? Why? If you can change something, what will it be ?

STEP UP FOR YOUR APPS :

1. Create a Q and A app as a review aid for the chapter's contents.

2. Create a point and click app of the timeline of digital electronics.

3. Think of an app that will be useful for you in relation to this chapter?

CHAPTER 2 BUILDING BLOCKS OF COMBINATIONAL LOGIC CIRCUITS

Learning Outcomes:

1. To identify and describe the basic logic gates.

2. To explain the symbols and operation of logic gates thru truth tables.

3. To identify and explain the universal gates.

4. To implement basic gates using universal gates.

5. To explain the process of making Integrated Circuits

6. To identify the different levels of Integration of Integrated Circuits.

Logic gates form the basic building blocks of Combinational Logic Circuits (CLC). Combinational Logic implements Boolean functions or logical operations mapping the relationship of the output variables to the input variables only. There are seven logical operations: OR, NOR, AND, NAND, NOT, Exclusive OR, and Exclusive NOR. The relationship of the output and the inputs are depicted in a Truth Table for all the possible combinations. For n inputs, there are 2^n combinations. For the logic gates below, let X and Y be the inputs and F be the output.

I – BASIC GATES

I -OR GATE

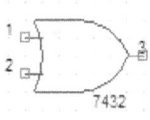

X	Y	F
0	0	0
0	1	1
1	0	1
1	1	1

A.1 Symbol　　　　　A.2 Truth Table

A.3 Equation

$F(X,Y) = X + Y$

For an OR gate, the output is 1 if and only if any of the inputs is 1.

II. AND GATE

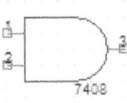

X	Y	F
0	0	0
0	1	0
1	0	0
1	1	1

B.1 Symbol

B.2 Truth table

B.3 Equation

$F(X,Y) = X \cdot Y$

For the AND gate, the output is 1 if and only if all inputs are 1.

III - THE INVERTER (THE NOT GATE)

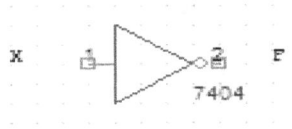

X	F
0	1
1	0

C.1 Symbol

C.2 Truth Table

C.3 Equation

$F = X'$

IV - NOR GATE

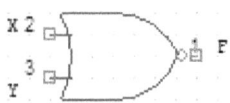

X	Y	F
0	0	1
0	1	0
1	0	0
1	1	0

D.1 Symbol   D.2 Truth Table

D.3 Equation

$F = (X + Y)'$

For the NOR gate, the output is 1 if and only if all of the inputs are zeroes

V - NAND GATE

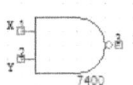

X	Y	F
0	0	1
0	1	1
1	0	1
1	1	0

F.1 Symbol   F.2 Truth Table

F.3 Equation

$F = (X.Y)'$

For the NAND gate, the output is 0 if and only if all of the inputs are zeros

VI - Exclusive – OR

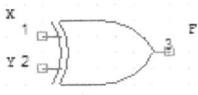

X	Y	F
0	0	0
0	1	1
1	0	1
1	1	0

F.1 Symbol   **F.2 Truth Table**

$F = X \text{ xor } Y$

F.3 Equation

For the exclusive or (xor), the output is one if the number of ones in the input is one.

II - THE UNIVERSAL GATES NAND AND NOR

For actual implementation, only one kind of gate is used to simplify the circuitry and for efficiency as well. The universal gates NAND or NOR is used to implement all the other gates. Thus, a logic diagram may be all NAND or all NOR. The basic gates NOT, AND and OR can be implemented with all NAND or all NOR gates.

A. THE UNIVERSAL GATE NOR

A.1 AND GATE IMPLEMENTED WITH NOR GATES

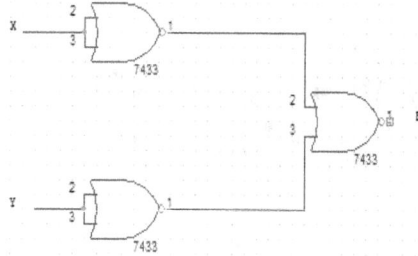

By De Morgan's :

$$F = (X' + Y')' = X'' \cdot Y'' = XY$$

A.2 OR GATE IMPLEMENTED WITH NOR

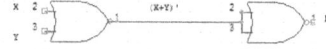

B. THE UNIVERSAL GATE NAND

B.1 AND GATE IMPLEMENTED WITH NAND

F = (XY)'' = XY

B.2 OR GATE IMPLEMENTED WITH NAND

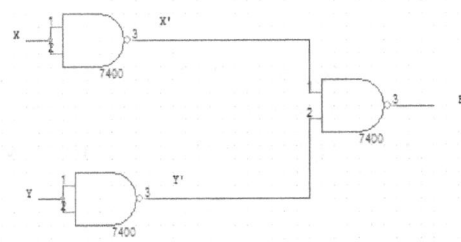

By De Morgan's:

(X' . Y')' = X'' + Y'' = X + Y

III - LEVELS OF INTEGRATION OF INTEGRATED CIRCUITS

As Gordon Moore has proposed , the number of components in an Integrated Circuit or chip doubles every two years . This complexity has led to different levels of integration . The level of integration defines the number of gates that can be placed inside a silicon chip. The different levels in the order of scale are the following :

1. Small Scale Integration (SSI). The number of gates is less than ten gates. Since the inputs and the outputs of the gates are connected directly to the to the pins of the IC the number of gates is limited by the number of pins in the device.

2. Medium Scale Integration (MSI) . The number of gates is between 10 to 100 in a single device. Basic functions are usually performed by this chip.

3. Large Scale Integration (LSI) . The number of gates is between 100 and several thousands in one chip. This sees applications in small memories and programmable modules.

4. Very Large Scale Integration (VLSI). The number of gates is between several thousands to million of gates in one chip. Microprocesor and microcomputer chips have started with this technology and allowed for the miniaturization of devices.

5 A complete timeline of the Levels of Integration up to date is vividly described in Intel.com. The company pioneered the manufacture of Integrated Circuits from the first chip as proposed by Robert Noyce to the chips that can be found in our computers, smartphones and many of our gadgets.

CHAPTER 2 END OF THE CHAPTER ACTIVITIES:

1. Generate a truth table for a three-input OR gate. How many possible combinations.

2. Generate the truth table for a four-input AND gate. How many possible combinations?

3. What follows after VLSI?

4. Give examples of SSI chips.

5. Give examples of MSI chips.

6. Give examples of LSI chips.

7. Give examples of VLSI chips.

8. What is TTL? How is the series of chips described? What is CMOS? How is the series of chips described?

LOOK UP - Online activities:

1. Look up for Logic gates materials and verify your current understanding.

2. Look up for video clips on actual chips and how they are wired.

3. Look up for the process of making an IC that won the Nobel prize for Physics.

4. Look up for the data sheets of the basic logic gates.

5. Look up for the different semiconductor companies manufacturing these chips.

PLANT VISIT : Arrange a visit to a local semiconductor company for a first hand glimpse of the process of making an IC. A glimpse of the segment of the process like assembly or testing is ok. Make a paper on your impressions of the trip.

REFLECTION PAPER:

Research on the different methods of making an IC. Compare this with the original method of making an IC that won the Nobel Prize for Physics. What were the key changes? Is Moore's law observable in the process as well?

STEP UP FOR YOUR APPS:

1. Make an interactive app about the Levels of Integration.

2. Make an interactive app about the process of making an IC.

3. Make an interactive app on the different logic gates and truth tables.

EXPERIMENT # 1: BASIC LOGIC GATES

Learning Outcomes:

1. To describe the actual chips of the basic logic gates committing to memory the pin configurations and the power requirements.

2. To interpret the data sheets of each chip.

3. To simulate the truth table of each chip.

Materials:

Digital Trainer or Breadboard, appropriate power supply (+5V), LEDs to monitor the output. Integrated Circuits (AND gate – 7408), (NAND gate – 7400), (INVERTER – 7404), (OR gate – 7432), (NOR gate -7433), (XOR gate – 7486), connecting wires. Data Sheets.

Procedure:

Step 1: From the data sheets, draw on the space below the configuration of the ICs.

AND gate – 7408

NAND gate – 7400

INVERTER – 7400

OR gate - 7432

NOR gate -7433

XOR gate – 7486

Step 2. Assemble the chips on the breadboard. Apply the appropriate supply voltage, Vcc = 5V, connect the ground pin to common ground. Connect a LED on the output pin in series with a resistor to monitor the status of the output. Complete the Truth Tables below by applying the indicated input signals and recording ON when the output LED is on and OFF when the output LED is off on the OUTPUT portion of the truth table. Indicate the pin numbers used. Refer to Step 1.

AND Gate

INPUT		OUTPUT
Pin# ___	Pin # ___	Pin # ___
Gnd	Gnd	
Gnd	5V	
5V	Gnd	
5V	5V	

NAND Gate

INPUT		OUTPUT
Pin# ___	Pin # ___	Pin # ___
Gnd	Gnd	
Gnd	5V	
5V	Gnd	
5V	5V	

OR Gate

INPUT		OUTPUT
Pin# ___	Pin # ___	Pin # ___
Gnd	Gnd	
Gnd	5V	
5V	Gnd	
5V	5V	

NOR Gate

INPUT		OUTPUT
Pin# ___	Pin # ___	Pin # ___
Gnd	Gnd	
Gnd	5V	
5V	Gnd	
5V	5V	

Ma. Beatriz Lacsamana

EXCLUSIVE OR Gate

INPUT		OUTPUT
Pin# ___	Pin # ___	Pin # ___
Gnd	Gnd	
Gnd	5V	
5V	Gnd	
5V	5V	

Step 3 : Observations

_____
_____
_____
_____
_____

Step 4 : Feedback (suggestions on how to make the experiment experience better)

_____
_____
_____
_____
_____

CONCLUSION:

CHAPTER 3 - REVIEW OF BOOLEAN ALGEBRA

Learning Outcomes :

1. To define Boolean Algebra , its postulates and theorems.

2. To explain the importance of Boolean function simplification in Digital Circuits.

3. To explain the concepts of Duality and Complements of Functions.

4. To apply the concepts to simplify Boolean expressions.

KEYWORDS

Boolean Algebra is the subarea of algebra in which the values of the variables are the truth values, usually denoted as 1 and 0 respectively. The concept was introduced by George Boole in 1913, the term "Boolean Algebra" suggested by Sheffer in 1913 .[14]

Boolean Algebra provides a basic logic operations on binary numbers 0,1. [15]

The Boolean Algebra that we will work with will operate on variables and logical operators ruled by postulates and axioms. The working variables can assume only the digital values of '0' and '1'. The following are examples of simple Boolean expressions :

$$F(X, Y) = X + Y' \qquad F(X,Y) = X'.Y$$

Variables X and Y in this function may take a value of '0' or '1' . The basic logical operators as shown in the example are the logical OR " + " , the logical AND ". " and the logical NOT " ' ". The Boolean expressions are governed by Edward V. Huntington's postulates and theorems as described in Table 3.1 . " *Postulates* are assumed to be true and we need not prove them. They provide the starting point of a theorem. A *theorem* is a proposition that can be deduced from postulates " [16]

An OBE Approach to Logic Circuits and Digital Design: Step by Step

Table 3.1 : Postulates and Theorems[16]

				Description
1.a	X + X = X	1.b	X . X = X	
2.a	X + 1 = 1	2.b	X . 0 = 0	
3.a	X + 0 = X	3.b	X . 1 = X	
4.a	X + X' = 1	4.b	X . X' = 0	
5.a	(X')' = X	5.b		Involution
6.a	X + Y = Y + X	6.b	X . Y = Y . X	Commutative
7.a	X + (Y + Z) = (X+Y) + Z	7.b	X . (Y . Z) = (X . Y) . Z	Associative
8.a	X (Y + Z) = X.Y + X.Z	8.b	X + YZ = (X + Y) (X + Z)	Distributive
9.a	(X + Y)' = X' . Y'	9.b	(X.Y)' = X' + Y'	De Morgan
10.a	X + X . Y = X	10.b	X . (X + Y) = X	Absorption
11.a	XY + X'Z + YZ = XY + X'Z			Consensus

PRINCIPLE OF DUALITY

The dual of an algebraic expression is formed by interchanging the AND and OR operations .The 0s are replaced by 1s and 0's by 1's. It further states that a Boolean expression is valid if we take the dual of the expression on both sides.

Illustrative Problem # 1: Take the dual of F (X, Y, Z) = XY'Z + X'YZ'

DUAL of F = (X + Y' + Z) (X' + Y + Z')

Illustrative Problem # 2 : F (W, X, Y, Z) = (W + X' + Y' + Z) (W' + X + Y + Z')

DUAL of F = WX'Y'Z + W'XYZ'

COMPLEMENT OF A FUNCTION

The Complement of a Function is formed in two steps :

Step 1 : Take the Dual of the Function

Step 2 : Complement all the variables

Illustrative Problem # 3 : Take the complement of $F(X, Y, Z) = XY'Z + X'YZ'$

Solution :

Step 1: Take the dual of F

DUAL of $F = (X + Y' + Z)(X' + Y + Z')$

Step 2 : Complement each variable

$F' = (X' + Y + Z')(X + Y' + Z)$

IP # 4 : Take the complement of $F(W, X, Y, Z) = (W + X' + Y' + Z)(W' + X + Y + Z')$

Solution :

Step 1 Take the dual of F

DUAL of $F = WX'Y'Z + W'XYZ'$

Step 2 : Complement each variable

$F' = W'XYZ' + WX'Y'Z$

SIMPLIFICATION OF BOOLEAN EXPRESSIONS

Simplification of Boolean Functions representing logic circuits is important. The goal is to come up with the simplest circuit with the minimum number of components. This is the cheapest and best implementation. To achieve this, we use our knowledge of postulates and theorems to manipulate a complex function into its simplest form.

IP #5: Simplify F(X, Y, Z) = X'YZ + XYZ + XY'Z + XY'Z'

$$= YZ (X' + X) + XY' (Z + Z') \qquad \text{Table 3.1 - 4.a}$$

ANS. F(X,Y,Z) = YZ + XY'

IP # 6 : Simplify F(X, Y, Z) = XYZ' + YZ' + X'YZ

Solution:

$$F = YZ' (X + 1) + X'YZ \qquad \text{Table 3.1 - 2.a}$$

ANS. F (X, Y, Z) = YZ' + X'YZ

IP # 7 : Given : F = (X + Y) (X + Z')

Solution :

$$F = XX + XZ' + XY + YZ' \qquad \text{Table 3.1 - 8.a}$$
$$= X + XZ' + XY + YZ'$$
$$= X (1 + Z' + Y) + YZ' \qquad \text{Table 3.1 - 2.a}$$

ANS. F (X, Y, Z) = X + YZ'

IP # 8 : Simplify : $F(X,Y,Z) = (X' + Y)(X + Y)$

Solution :

$$F = X'X + X'Y + XY + YY$$
$$= 0 + Y(X' + X) + YY$$
$$= 0 + Y + YY$$
$$= Y(1 + Y)$$

ANS. $F(X,Y,Z) = Y$

IP # 9 : Prove the Absorption Theorem $X \cdot (X + Y) = X$

Solution:

$XX + XY = X$	Table 3.1- 8.a
$X + XY = X$	Table 3.1-1.b
$X(1 + Y) = X$	Table 3.1- 2.a
$X = X$	

IP # 10 : Prove the Consensus Theorem $XY + X'Z + YZ = XY + X'Z$

Solution :

$XY + X'Z + YZ = XY + X'Z$	
$XY + X'Z + YZ(X + X') = XY + X'Z$	Table 3.1 - 4.a
$XY + X'Z + XYZ + X'YZ = XY + X'Z$	Table 3.1 - 8.a
$XY(1 + Z) + X'Z(1 + Y) = XY + X'Z$	Table 3.1 - 2.a

APPLICATION :

IP # 11 : Given the logic diagram shown below and described by the following function:

$$F(X, Y, Z) = XY'Z' + XY'Z + Y'Z$$

Logic Diagram :

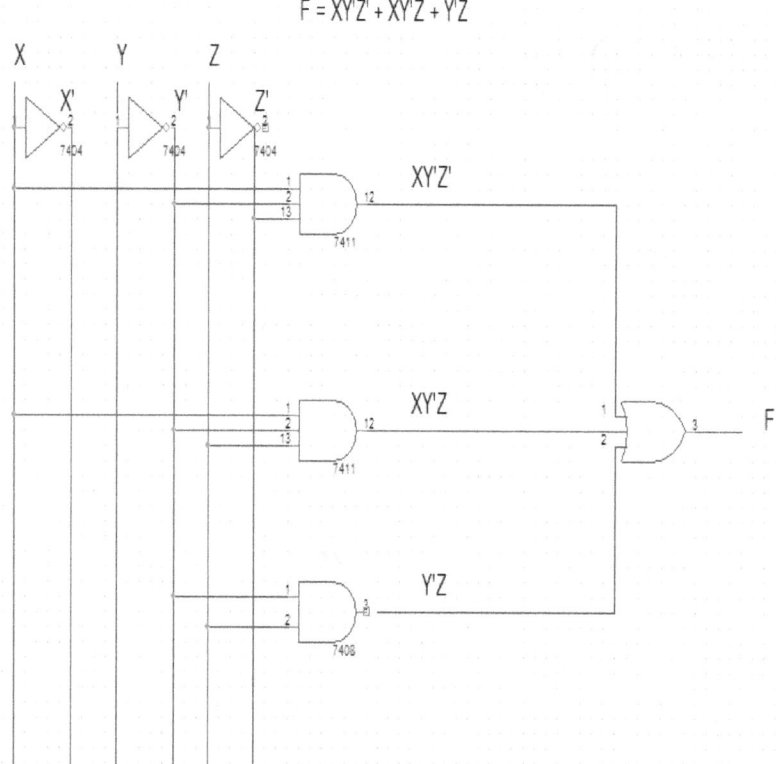

Compare the Cost in terms of components and inputs saved by simplifying the function.

Solution :

$$F = XY'Z' + Y'Z (X + 1)$$
$$= XY'Z' + Y'Z$$

Simplified Logic Diagram:

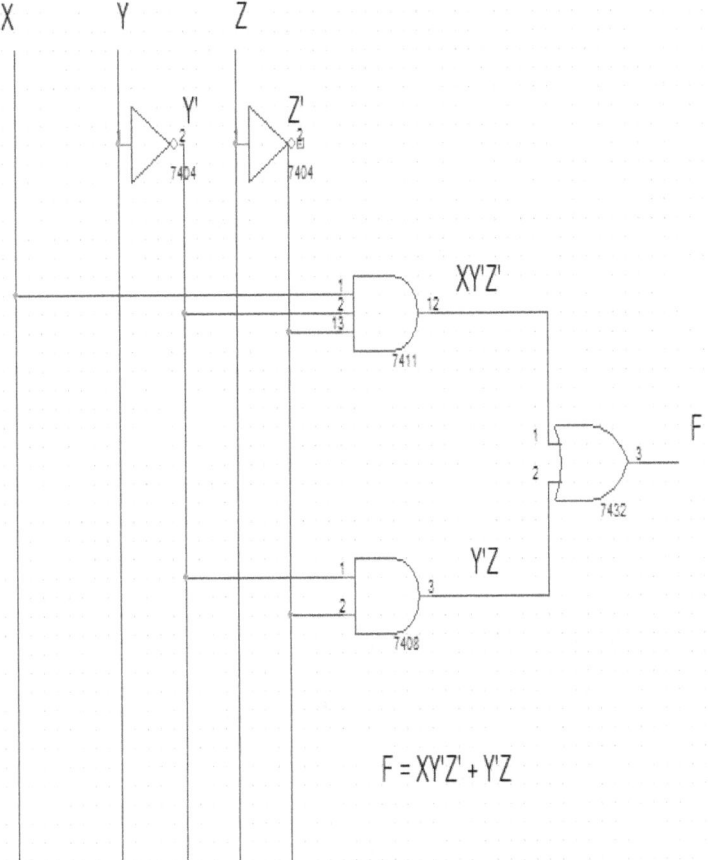

$$F = XY'Z' + Y'Z$$

Table 3.2 - Cost table

Original Function		Simplified Function	
Number of Inputs	Number of Gates	Number of Inputs	Number of Gates
5	7	5	5
XYZ		XYZ	
Y'Z'		Y'Z'	

SAVINGS : 2 gates

CHAPTER 3 END OF THE CHAPTER ACTIVITIES:

1. Draw the simplified logic diagram for Illustrative Problems 1 to 10 and generate the Cost Table.

2. Simplify the following functions :

   a. $F(X, Y, Z) = XYZ' + X'YZ + YZ$

   b. $F(X,Y,Z) = XY + X'Z + YZ$

   c. $F(X,Y,Z) = (X' + Y)(X + Y)$

   d. $F(X,Y,Z) = (X + Y')(X + Y)$

   e. $F = (X + Y)(X' + Z)$

3. Take the DUAL of the simplified functions in number 2.

4. Take the Complement of the simplified functions in number 2.

5. Draw the Logic diagram of the simplified functions in number 2.

6. Draw the Logic Diagram of the functions in number 4.

EXPERIMENT 2 : SIMPLIFICATION OF A FUNCTION USING BOOLEAN ALGEBRA

Learning Outcomes:

1. To simulate the output of a function and its simplified form.

2. To recognize that the output of the two logic circuits are the same.

3. To recognize how much is saved through the use of a COST Table.

Materials :

Digital Trainer or Breadboard, appropriate power supply (+5V), LEDs to monitor the output. Integrated Circuits (two input AND gate – 7408), (three input AND gate – 7411) (two input OR gate – 7432), (three input OR gate), connecting wires. Data Sheets.

Experimental Circuit :

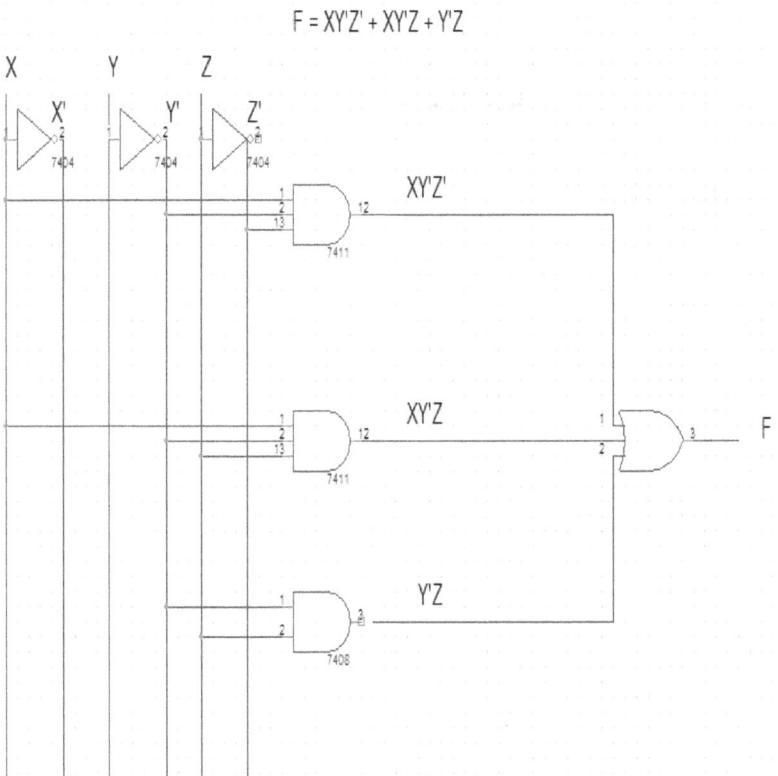

$F = XY'Z' + XY'Z + Y'Z$

Step 1 : Wire on the breadboard the experimental circuit as shown

Step 2 : Fill up the OUTPUT of the Truth Table below. Indicate ON if the output LED light is ON and OFF otherwise on column F.

INPUT			OUTPUT
X	Y	Z	F
Gnd	Gnd	Gnd	
Gnd	Gnd	5V	
Gnd	5V	Gnd	
Gnd	5V	5V	
5V	Gnd	Gnd	
5V	Gnd	5V	
5V	5V	Gnd	
5V	5V	5V	

Step 3: Simplify the Boolean function describing the experimental circuit.

Step 4 : Draw the simplified circuit .

Step 5 : Generate the new Truth Table .

Step 6: Compare the two circuits by generating the cost tables for each one below. How much is saved in the simplified circuit.

CONCLUSION

CHAPTER 4 : MULTILEVEL COMBINATIONAL LOGIC CIRCUITS

Learning Outcomes:

1. To describe a multilevel Combinational Logic Circuit.

2. To generate the output of a multilevel CLC point by point.

3. To generate the Truth Table of a MCLC.

4. To convert circuits to the universal gates, NAND and NOR.

5. To diagram and contrast universal NAND and universal NOR circuits

Multilevel Combinational Logic Circuits (MCLC) are made up of several logic gates which are cascaded to form an output. Between the input and the output are intermediate points. Like basic logic gates the behavior of MCLCs are best described using truth tables. The difference is that the truth table of an MCLC includes the behavior of the circuit on the Intermediate Points. In addition to the input and output, the Intermediate Points are entered into the Truth Table.

Illustrative Problem #1 : MCLC with three Intermediate Points: A, B, C. The output is at point D. Generate the equations for each Intermediate Point and the Truth table for the following circuit.

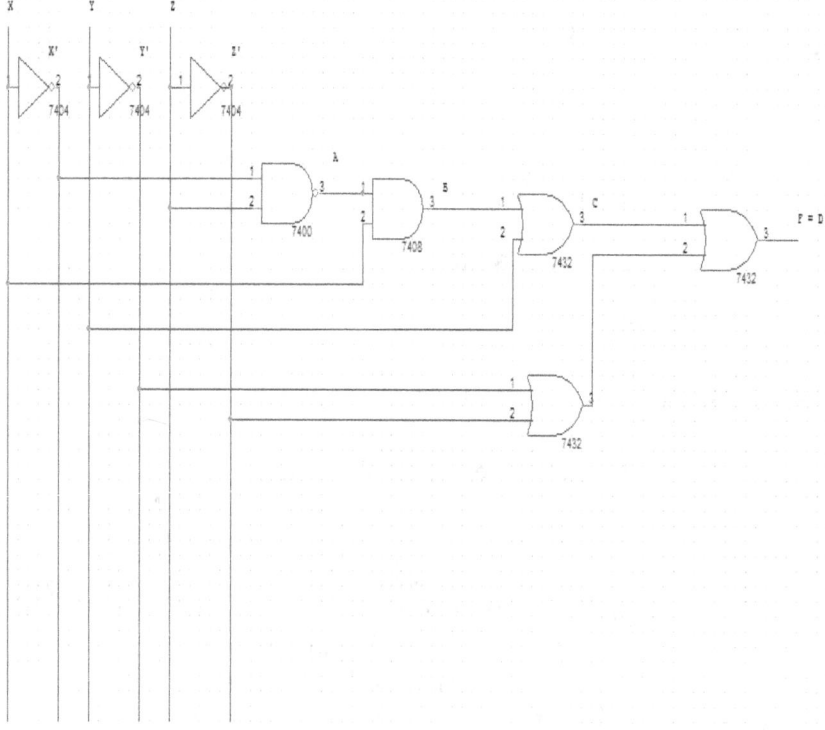

Step 1 : Equations for the Intermediate Points of the Circuit

A (X, Y, Z) = (X'Z)'

B = A . X

C = B + Y

F = D = C + (Y' + Z')

Step 2 : Generate the Truth Table. Since there are three inputs, there will be 8 possible combinations. In addition to the three inputs, there are three Intermediate Points, A, B, C and one output D.

X	Y	Z	A	B	C	F = D
0	0	0	1	0	0	1
0	0	1	0	0	0	1
0	1	0	1	0	1	1
0	1	1	0	0	1	1
1	0	0	1	1	1	1
1	0	1	1	1	1	1
1	1	0	1	1	1	1
1	1	1	1	1	1	1

Illustrative Problem # 2 : MCLC with four Intermediate Points A, B, C, D and four inputs W,X,Y,Z. There are also two outputs F1 and F2.

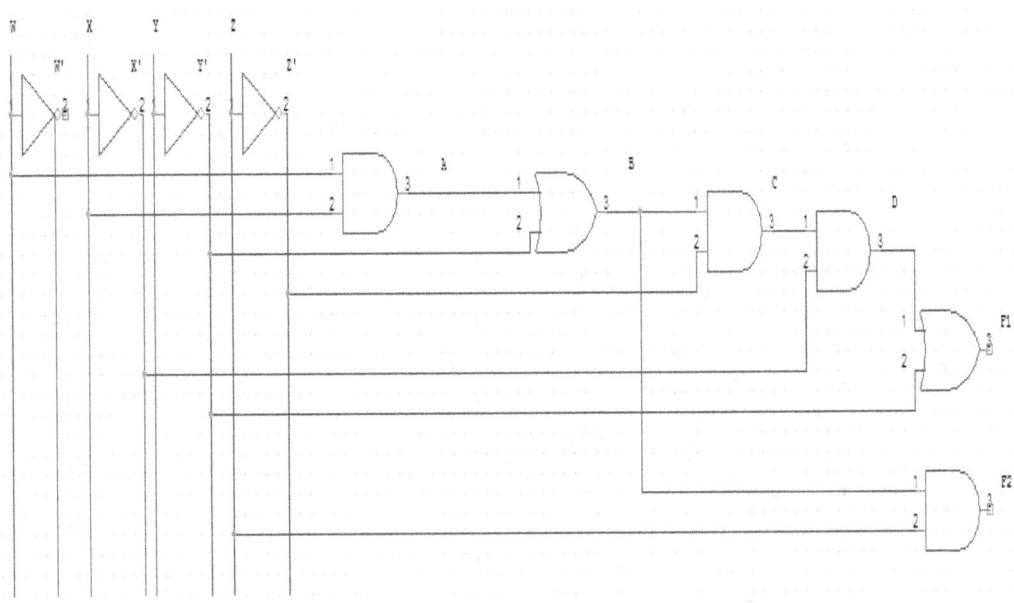

Step 1 : Generate the equations on each Intermediate Point

$A(W,X,Y,Z) = WX$

$B = A + Y'$

$C = B \cdot Z'$

$D = C \cdot X'$

$F1 = D + Y'$

$F2 = B \cdot Y'$

An OBE Approach to Logic Circuits and Digital Design: Step by Step

Step 2 : Generate the Truth Table . Since there are four inputs W,X,Y,Z, there are 16 possible combinations. There are four intermediate points A, B, C, D and two outputs F1 and F2.

W	X	Y	Z	A	B	C	D	F1	F2
0	0	0	0	0	1	1	1	1	1
0	0	0	1	0	1	0	0	1	1
0	0	1	0	0	0	0	0	0	0
0	0	1	1	0	0	0	0	0	0
0	1	0	0	0	1	0	0	1	1
0	1	0	1	0	1	1	0	1	1
0	1	1	0	0	0	0	0	0	0
0	1	1	1	0	0	0	0	0	0
1	0	0	0	0	1	1	1	1	1
1	0	0	1	0	1	0	0	1	1
1	0	1	0	0	0	0	0	0	0
1	0	1	1	0	0	0	0	0	0
1	1	0	0	1	1	1	0	1	1
1	1	0	1	1	1	0	0	1	1
1	1	1	0	1	1	1	0	0	0
1	1	1	1	1	1	0	0	0	0

The examples above illustrated that going through the intermediate points facilitated the generation of the output. It can also be observed that multilevel circuits tend to be cumbersome that it is why in the next chapters a methodology is introduced so as to reduce MCLCs into standard two-level circuits.

Illustrative Problem # 3 : Convert the following circuit to an all NAND circuit

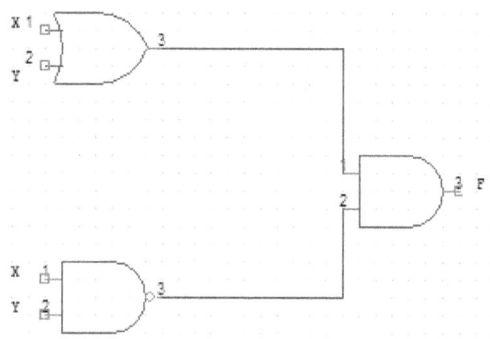

ANSWER :

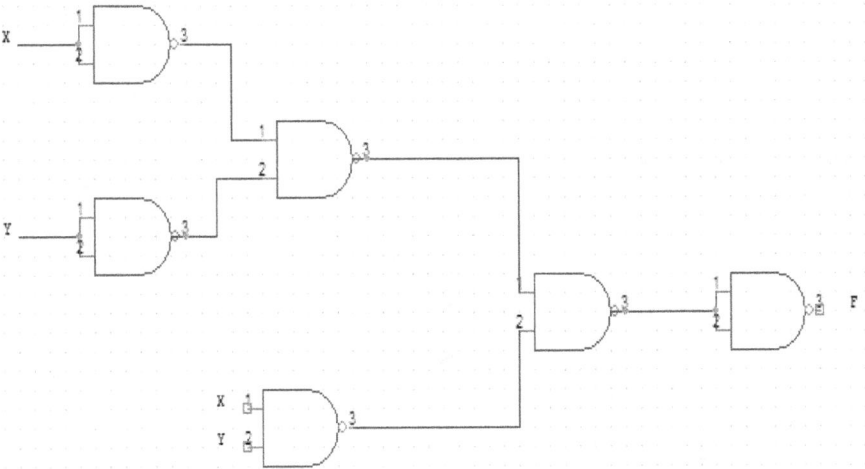

An OBE Approach to Logic Circuits and Digital Design: Step by Step

Illustrative Problem # 4 : Convert the following circuit into an all NOR circuit

ANSWER :

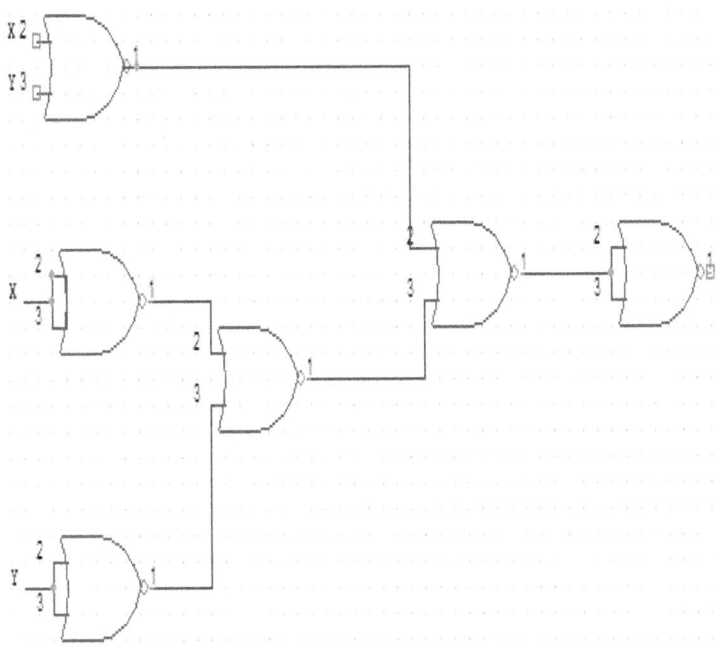

The examples above have demonstrated that circuits can be converted to an all NAND or all NOR circuit.

CHAPTER 4 END OF THE CHAPTER ACTIVITIES

Construct the Truth table for the following MCLS by going through each Intermediate Point until you reach the output.

1. Four Level MCLC with two outputs

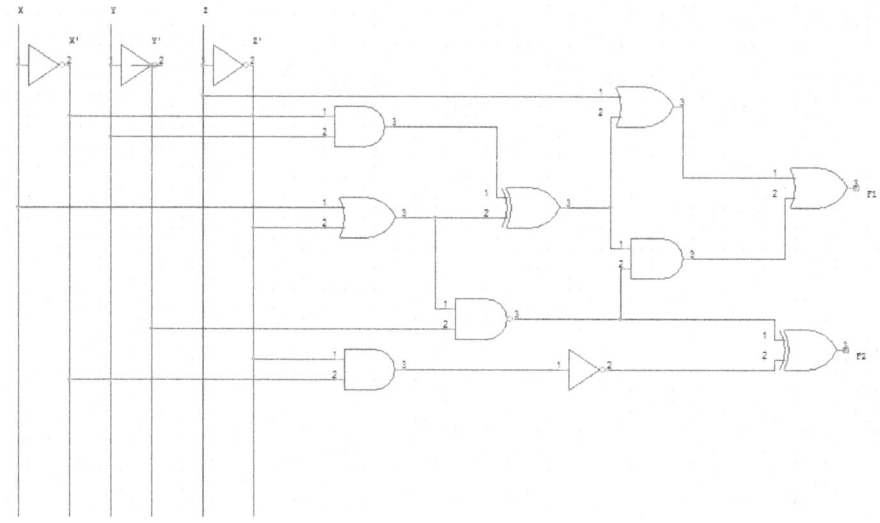

2. Three level MCLC with three outputs.

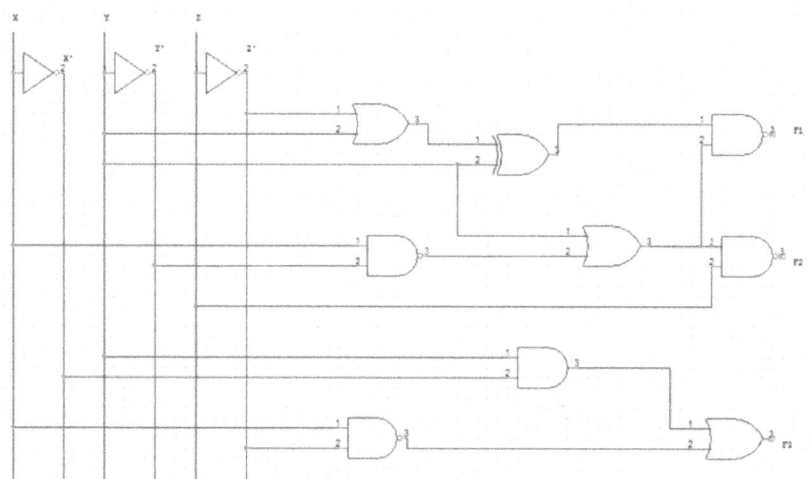

3. Three level MCLC with two outputs.

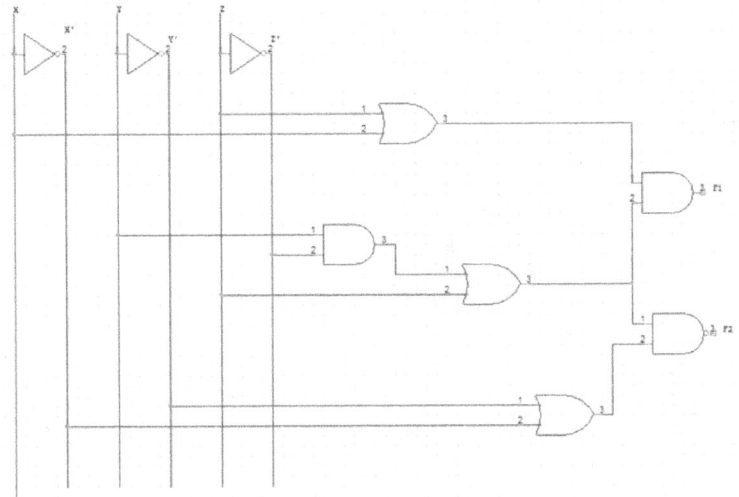

4. Two level MCLC with three outputs

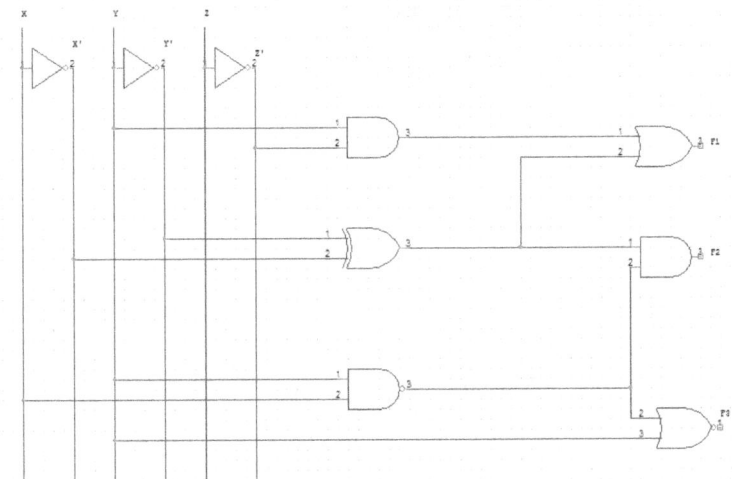

5. Three levels MCLC with two outputs

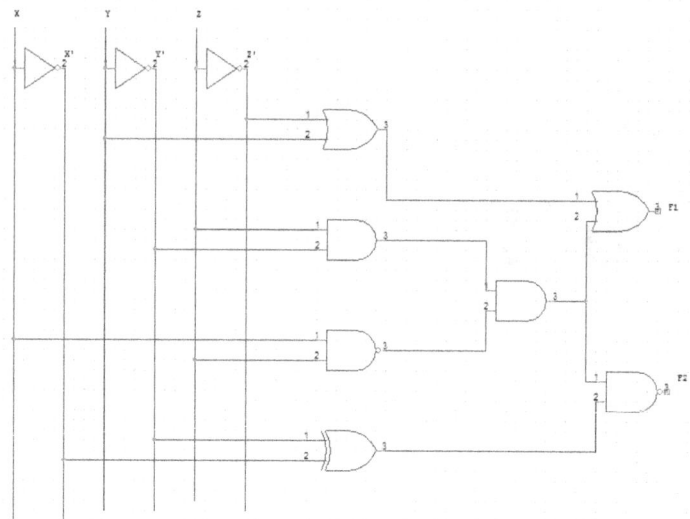

6. Convert Circuits 1-5 to an ALL NAND Circuit

7. Convert Circuits 1-5 to an ALL NOR Circuit

EXPERIMENT 3 : UNIVERSAL GATES

Learning Outcomes :

1. To demonstrate how logic circuits constructed form basic gates can be converted to an ALL –NAND or and ALL –NOR circuits.

2. To simulate and compare the outputs of these two circuits.

Materials :

Digital Trainer or Breadboard, appropriate power supply (+5V) , LEDs to monitor the output. Integrated Circuits (two input AND gate – 7408), (three input AND gate – 7411) (two input OR gate – 7432), (three input OR gate), connecting wires. Data Sheets.

Procedure:

Step 1 : Wire the experimental circuit below :

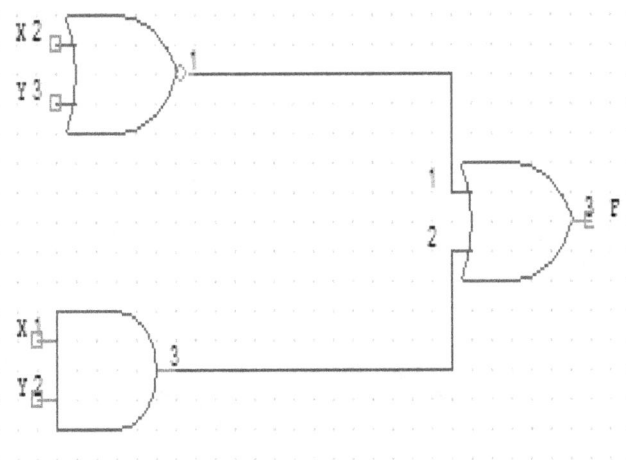

Step 2 : Generate the Truth Table

Step 3 : Convert to an ALL NAND circuit.

3.a Draw the new Experimental Circuit

3.b Generate the Truth Table

Step 4. Convert to an ALL NOR circuit.

4.a Draw the new experimental circuit.

4.b Generate the Truth Table.

Step 5. Compare the Truth Tables of Steps 2, 3.b and 4.b. Are they all the same ?

CONCLUSION

_____
_____
_____

CHAPTER 5 - STANDARD FORMS and TWO-LEVEL IMPLEMENTATION

Learning Outcomes :

1. To demonstrate how to express Boolean expression in Standard Form.

2. To explain why two-level circuits are better than MCLCs.

3. To describe how to convert from one Standard Form to the other.

4. To explain the value of converting Boolean expressions in standard form.

KEYWORDS

Product Term - a term with variables logically ANDed.

Minterm - a product term where all the input variables are logically ANDed.

Sum Term - a term with logical ORed variables

Maxterm - a sum term where all the input variables are logically ORed.

Sum of Products (SOP) - a Boolean expression where product terms are logically ORed.

Product of Sums (POS) - a Boolean expression where sum terms are logically ANDed.

Standard Form - refers to Sum of Products (SOP) and Product of Sums (POS)

Canonical Form - refers to the Sum of Minterms (SOM) and the Product of Maxterms (POM)

Boolean expressions are simplified into a two-level implementation called a standard form. There are two types: the Sum of Products (SOP) and the Product of Sums (POS) . When all the input variables are present in each product term, we have a minterm. Thus, this SOP is called a SOM or Sum of Minterms. The same is true for Product of Sums, if all the variables are present in the sum term, we have a Maxterm. The POS is called a Product of Maxterm.

SUM OF PRODUCTS (SOP) OR SUM OF MINTERMS (SOM)

The Sum of Products is the logical OR of product terms. It is an "ORing Process". For example, F = XYZ + X'YZ + XY'Z. The minterms or product terms in this expression are XYZ, X'YZ and XY'Z. It is the logical AND of all the input variables. The true value for a product term is 1. This means that all variables in a product term must be a 1 or be complemented to become 1. The normal value of product term is 1. The minterms are best explained using the following table:

Table 5.1 Minterms Table for three variables (X,Y,Z)

index	X	Y	Z	minterms
0	0	0	0	X'Y'Z'
1	0	0	1	X'Y'Z
2	0	1	0	X'YZ'
3	0	1	1	X'YZ
4	1	0	0	XY'Z'
5	1	0	1	XY'Z
6	1	1	0	XYZ'
7	1	1	1	XYZ

Illustrative Problem 1 : Abbreviate the following SOM expression: F(X,Y,Z) = XYZ + X'YZ + XY'Z

Solution : Refer to the table and search for the index equivalence, which is the decimal equivalent of the binary representation of the minterm.

Answer : $F(X,Y,Z) = \sum_m (7,3,6)$

    Reorder in ascending sequence:

    $F(X,Y,Z) = \sum_m (3, 6, 7)$

        " Sum of minterms 3, 6 and 7"

Notes : $\sum_m$ - reads "sum of minterms"

SHORTCUT : For this example, since there are three input variables, recall the binary positional value for 3 bits which is 4 2 1. From there, get the equivalent value of each term. For example, X'YZ corresponds to 011 or 3.

Table 5. 2 Function Table

X	Y	Z	minterms		m0	m1	m2	m3	m4	m5	m6	m7
0	0	0	X'Y'Z'	m0	1							
0	0	1	X'Y'Z	m1		1						
0	1	0	X'YZ'	m2			1					
0	1	1	X'YZ	m3				1				
1	0	0	XY'Z'	m4					1			
1	0	1	XY'Z	m5						1		
1	1	0	XYZ'	m6							1	
1	1	1	XYZ	m7								1

Illustrative Problem # 2 : Rewrite the abbreviated form in IP#1 in terms of the designated symbol for each minterm.

Answer : $F(X,Y,Z) = m3 + m6 + m7$

Illustrative Problem # 3 : Given : F(X,Y,Z) = X'Y'Z' + X'YZ' + XY'Z + XYZ

1. Express in abbreviated form.

2. Express in terms of designated minterm symbol.

Answer:

1 : $F(X,Y,Z) = \sum m\,(0,2,5,7)$

2. $F(X,Y,Z) = m0 + m2 + m5 + m7$

From Table 5.2, it can be seen that for n input variables there are 2^n minterms. The minterms are designated as m_i where i is the value when the term is equal to 1. For this table the minterms are X'Y'Z', X'Y'Z, X'YZ', X'YZ, XY'Z', XY'Z, XYZ', XYZ corresponding to m0, m1, m2, m3, m4, m5, m6 and m7 respectively. The true value for minterms is 1 and the complemented value is zero. Thus all variables assuming the value of zero is complemented. For example in m0 where X, Y, Z are all zeroes, the minterm will have all variables complemented as m0 = X'Y'Z'.

Table 5. Function Table for IP#3

i	X	Y	Z	F	m0	m2	m5	m7	m0 + m2 + m5 + m7
0	0	0	0	1	1	0	0	0	1
1	0	0	1	0	0	0	0	0	0
2	0	1	0	1	0	1	0	0	1
3	0	1	1	0	0	0	0	0	0
4	1	0	0	0	0	0	0	0	0
5	1	0	1	1	0	0	1	0	1
6	1	1	0	0	0	0	0	0	0
7	1	1	1	1	0	0	0	1	1

From Table 5, we can clearly see that F = m0 + m2 + m5 + m7. This is logical since F = 1, only at input combinations i= 0,2,5 and 7. Thus, by OR - ring minterm m0 (which has a value of 1 only at input combination i=0), with minterm m2 (which has a value of 1 only at input combination i= 2), with minterm m5 (which has a value of 1 only at input combination i= 5) with minterm m7 (which has a value of 1 only at input combination i= 7) the resulting function will be equal to F. Note also that the presence of a 1 for each minterm in the table ensured that F= 1 which is true for a logical OR function which states that if any of the input is 1 the output is 1. Therefore, any function can be expressed by OR-ing all minterms (mi) corresponding to input combinations (i) at which the function has a value of 1. The resulting expression is commonly known as the SUM of Minterms and is typically expressed as F = $\sum_m(0, 2, 5, 7)$, where $\sum$ indicates OR-ing of the indicated minterms. Thus, F = $\sum_m(2, 4, 5, 7)$ = (m0+ m2 + m5 + m7)

PRODUCT OF SUMS (POS) or PRODUCT OF MAXTERMS (POM)

The Product of Maxterms is the logical AND of Maxterms or sum terms. It is a "logical AND Process". For example, $F = (X + Y + Z)(X' + Y + Z)(X + Y' + Z)$. The Maxterms in this expression are $X + Y + Z$, $X' + Y + Z$ and $X + Y' + Z$. It is the logical OR of all the input variables. The true value for a sum term is 0. This means that all variables in a sum term must be a 0 or be complemented to become 0. The Maxterms are best explained using Table 5.3.

Table 5.3 Maxterms Table for three variables (X,Y,Z)

index	X	Y	Z	Maxterms
0	0	0	0	$X + Y + Z$
1	0	0	1	$X + Y + Z'$
2	0	1	0	$X + Y' + Z$
3	0	1	1	$X + Y' + Z'$
4	1	0	0	$X' + Y + Z$
5	1	0	1	$X' + Y + Z'$
6	1	1	0	$X' + Y' + Z$
7	1	1	1	$X' + Y' + Z'$

Illustrative Problem 4: Abbreviate the following POM expression:

$F(X,Y,Z) = (X+Y+Z)(X'+Y+Z)(X+Y'+Z)$

Solution: Refer to the table and search for the index equivalence, which is the decimal equivalent of the binary representation of the Maxterm.

Answer: $F(X,Y,Z) = \prod_M (7,3,6)$

Reorder in ascending sequence:

$F(X,Y,Z) = \prod_M (3, 6, 7)$

" Product of Maxterms 3, 6 and 7"

Notes $\prod_M$ reads "Product of Maxterms"

Illustrative Problem # 5 : Abbreviate the following POM expression

Given : $F(X,Y,Z) = (X + Y + Z)(X' + Y + Z')(X + Y' + Z)(X' + Y' Z')$

1. Express in abbreviated form.

2. Express in terms of designated Maxterm symbol.

Answer:

1 : $F(X,Y,Z) = \prod_M (0,2,5,7)$

2. $F(X,Y,Z) = M0 + M2 + M5 + M7$

Table 5. 4 Function Table

X	Y	Z	Maxterms		Mo	M1	M2	M3	M4	M5	M6	M7
0	0	0	X + Y + Z	M0	0							
0	0	1	X + Y + Z'	M1		0						
0	1	0	X + Y' + Z	M2			0					
0	1	1	X + Y' + Z'	M3				0				
1	0	0	X' + Y + Z	M4					0			
1	0	1	X' + Y + Z	M5						0		
1	1	0	X' + Y' + Z	M6							0	
1	1	1	X' + Y' + Z'	M7								0

From Table 5.4 it can be seen that for n input variables there are 2^n maxterms. The maxterms are designated as M_j where j is the value when the term is equal to 0. For this table the Maxterms are (X + Y + Z), (X + Y + Z'), (X + Y' + Z), (X + Y' + Z'), (X' + Y + Z), (X' + Y + Z'), (X' + Y' + Z) and (X' + Y' + Z') corresponding to M0, M1, M2, M3, M4, M5, M6 and M7 respectively. The true value for Maxterms is 0 and the complemented value is 1. Thus all variables with the value of one is complemented. For example in M7 where X, Y, Z are all ones, the Maxterm will have all variables complemented as M7 = (X' + Y' + Z').

Table 6. Function Table for IP # 5

i	X	Y	Z	F	M0	M2	M5	M7	M0.M2.M5.M7
0	0	0	0	0	0	1	1	1	0
1	0	0	1	1	1	1	1	1	1
2	0	1	0	0	1	0	1	1	0
3	0	1	1	1	1	1	1	1	1
4	1	0	0	1	1	1	1	1	1
5	1	0	1	0	1	1	1	0	0
6	1	1	0	1	1	1	1	1	1
7	1	1	1	0	0	0	0	1	0

Relationship between Maxterms and minterm:

$$m_i = M_i'$$

Recall De Morgan's Theorem, which states that :

$(X + Y)' = X' + Y'$

$(X'Y')' = X'' + Y'' = X + Y$

PROCEDURE FOR CONVERSION BETWEEN FORMS:

Step 1 : Take the Complement

Step 2 : Change Form

Conversion between Sum of Products (SOP) and Product of Sums (POS)

Illustrative Problem # 6 : Convert the SOP expression, $F(X,Y,Z) = \sum m (1,3,5,7)$ to POS

Step 1 : Take the complement

$F'(X, Y, Z) = \sum_m (0,2,4,6)$

Step 2 : Change the form ($\sum_m$ to $\prod_M$)

$F(X, Y, Z) = \prod_M (0,2,4,6)$

Thus,

$$m_i = M_i'$$

Illustrative Problem # 7: Convert the POS expression, $F(W, X, Y, Z) = \prod_M (2,4,6,8,10,12,14)$

Step 1: Take the Complement

$F'(W,X,Y,Z) = (0,1,3,5,7,9,11,13,15)$

Step 2: Change the form ($\prod_M$ to $\sum_m$)

$F(W,X,Y,Z) = \sum m (2,4,8,10,12,14)$

Two-Level Implementations of Standard Forms

Sum of Products Expression (SOP):

An SOP expression can be implemented in 2-levels of gates. The first level consists of a number of AND gates which equals the number of product terms in the expression. Each AND gate implements one of the product terms in the expression. The second level consists of a SINGLE OR gate whose number of inputs equals the number of product terms in the expression.

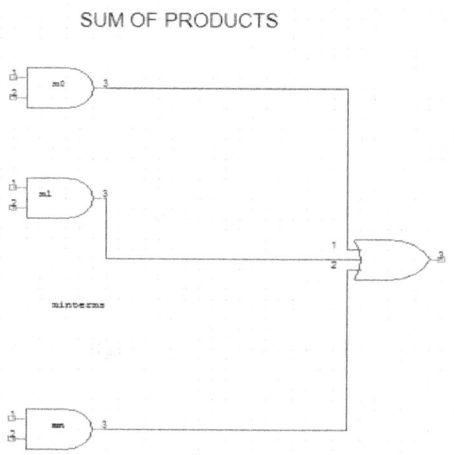

Illustrative Problem # 8 : Implement the SOP F(W,X,Y,Z) = WX' + W'XY'Z + YZ'

The circuit as shown consists of four inverters three AND gates and 1 OR gate in a two level implementation arrangement. Level 1 consists of the AND gates representing the product terms. The second level is the OR gate responsible for the OR-ing process.

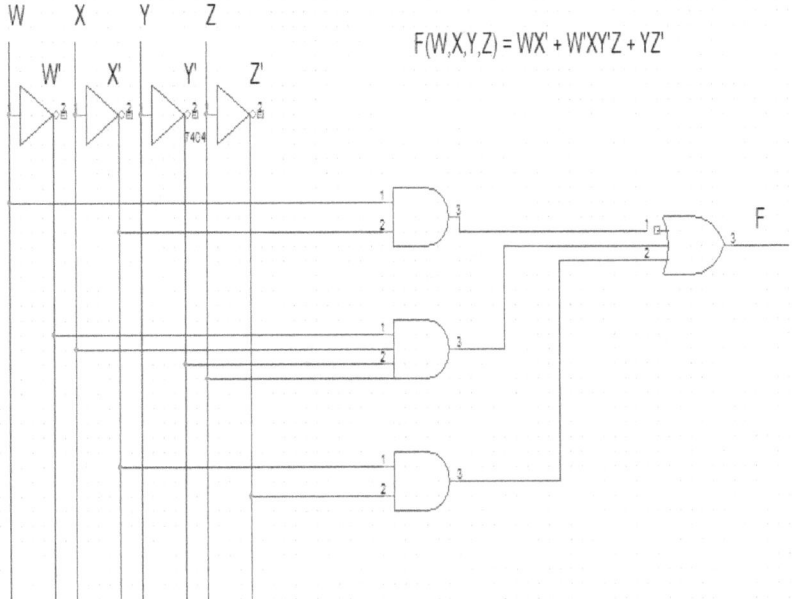

Product of Sums Expression (POS);

A POS expression can be implemented in 2-levels of gates. The first level consists of a number of OR gates which equals the number of sum terms in the expression. Each OR gate implements one of the sum terms in the expression. The second level consists of a SINGLE AND gate whose number of inputs equals the number of sum terms in the expression.

Illustrative Problem # 9 : Implement POS F(W Y,Z) =(W' + X) (W + X' + Y + Z') (Y' + Z)

The circuit as shown consists of four inverters, three OR gates and 1 AND gate in a two level implementation arrangement. Level 1 consists of the OR gates and representing the SUM terms. The second level is the AND gate responsible for the AND-ing process.

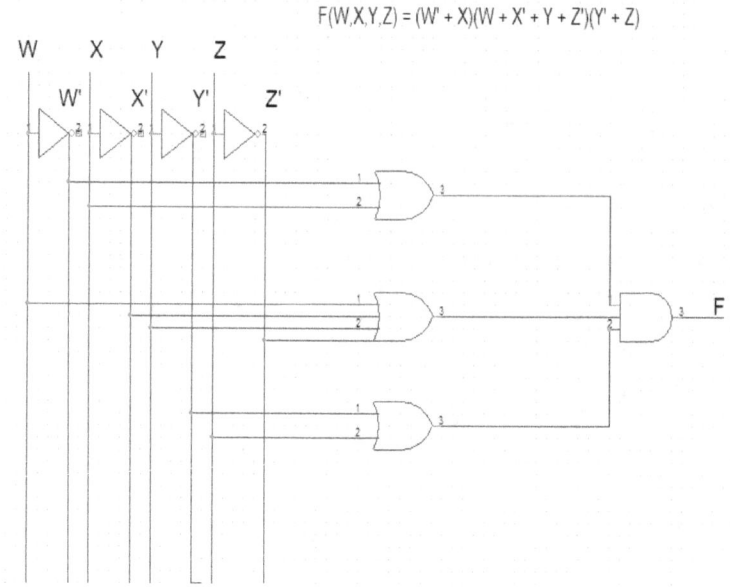

Illustrative Problem # 10 : Comparison of IP#8 and IP#9

A close examination of IP#8 and IP#9 will lead to the conclusion that they are complements of each other.

Each minterm, mi has a corresponding Mj'.

IP#8 F(W,X,Y,Z) = WX' + W'XY'Z + YZ'   IP#9 F(W,X, Y,Z) =(W' + X) (W + X' + Y + Z') (Y' + Z)

Illustrative Problem # 10 : F(X, Y, Z) = X'Y + Y'Z + XZ'

1. Convert the given SOP Expression above to SOM. Recall that in Sum of Minterms, each minterm contains all the input variables.

2. Convert to abbreviated form.

X'Y (Z + Z')   : Introduce Z. Recall that Z + Z' = 1

Y'Z (X + X')   : Introduce X. Recall that X + X' = 1

XZ' (Y + Y')   : Introduce Y . Recall that Y + Y' = 1

Therefore the equivalent SOM expression is :

$F(X, Y, Z) = X'YZ + X'YZ' + XY'Z + X'Y'Z + XYZ' + XY'Z'$

In abbreviated form,

$F(X, Y, Z) = \sum_m (3,2,5,1,6,4)$

Rearranging in ascending order :

$F(X, Y, Z) = \sum_m (1,2,3,4,5,6)$

CHAPTER 5 END OF THE CHAPTER ACTIVITIES :

1. What is the difference between Sum of Products and Sum of minterms?

2. What is the difference between Product of Sums and Product of Maxterms?

3. Convert to POS the following SOP Expressions. Draw the equivalent logic diagram for each one.

$F(X, Y, Z) = \sum_m (1,3,5,7)$

    a. $F(X, Y, Z) = \sum_m (0,2,4,6)$

    b. $F(X, Y, Z) = \sum_m (0,2,6,7)$

    c. $F(X, Y, Z) = \sum_m (2,4,5)$

    d. $F(X, Y, Z) = \sum_m (0,1,6,7)$

4. Convert to SOP the following POS Expressions. Draw the equivalent diagram for each one.

    a. $F(X, Y, Z) = \Pi_M (1,3,5,7)$

    b. $F(X, Y, Z) = \Pi_M (0,2,4,6)$

    c. $F(X, Y, Z) = \Pi_M (0,2,6,7)$

    d. $F(X, Y, Z) = \Pi_M (2, 4, 5)$

    e. $F(X, Y, Z) = \Pi_M (0, 1, 6, 7)$

6. LOOK UP – Online Activities :

1. Look up for more Boolean Expressions and simplify them
2. Look up for books and ebooks for further reading and exercise

7. STEP UP FOR YOUR APPS

1. Make an app that will convert SOP functions to POS functions
2. Make an app that will convert POS functions to SOP functions

EXPERIMENT 4 - STANDARD SOP TO POS: CONVERSION BETWEEN FORMS

Learning Outcomes :

1. To simulate and observe the actual behavior of Standard Form circuits.

2. To compare and determine the similarities and differences of an SOP and POS Equivalent circuits.

3. To compare Function tables of SOP and POS Circuits

Materials:

Digital Trainer or Breadboard, appropriate power supply (+5V) , LEDs and series resistor to monitor the output. Integrated Circuits (AND gate – 7408), (INVERTER – 7404), (OR gate – 7432), connecting wires. Data Sheets.

Procedure :

Step 1: Wire the SOP Experimental Circuit Below :

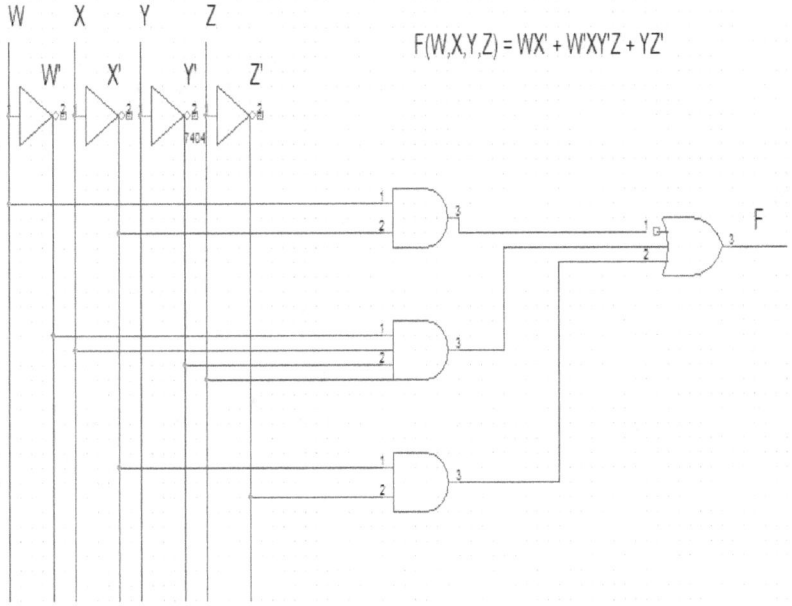

$F(W,X,Y,Z) = WX' + W'XY'Z + YZ'$

Step 2: Generate the Function Table from the inputs and outputs of the actual circuit. Make sure that your actual output is the same as the theoretical function table.

Step 3 : Derive the Equivalent POS Equation and Draw the circuit.

3.a POS Equation : _____

3.b Draw the POS Circuit

Step 4 : Generate the Function Table from the Experimental POS Circuit.

Step 5 : Compare the SOP and POS Equations and their Corresponding Function Tables

CONCLUSION

_____
_____
_____
_____

CHAPTER 6 – MINIMIZATION USING KARNAUGH MAPS

Learning outcomes :

1. To minimize Boolean expressions using K-maps.

2. To perform the process of using K-maps for 2-Variable, 3-Variable and 4-Variable Boolean expressions.

A Karnaugh map is an alternate way of representing a truth table. It is a systematic tool used to minimize Boolean expressions. For this chapter we will explore K-map minimization for 2, 3 and 4 variables only. For 5 variables and up, another method called the Quine-Mckluskey algorithm is best employed.

2- VARIABLE K-MAPS

For Boolean expressions of two variable n there are four product terms or minterms corresponding to four squares in the Karnaugh map as shown in Figure 6.1. A minterm present in the expression is represented as one in the map. The procedure for k-mapping is as follows:

Step 1 : Draw the K-map corresponding to the number of variables.

Step 2 : Enter "1 " for each given minterms on the appropriate square on the K-map.

Step 3: Group adjacent minterms. Two minterms are adjacent if only one variable changed between them. Grouping must correspond only to a power of two, 2^n

Step 4: The number of variables in the simplified equation is determined by the number of mintems in a group . For example :

$$4 \text{ Minterms} = \text{``1''}$$

$$2 \text{ Minterms} = 1 \text{ Variable}$$

$$1 \text{ Minterm} = 2 \text{ Variables}$$

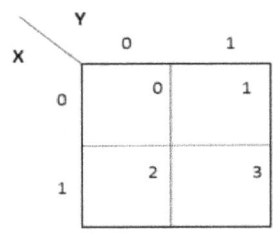

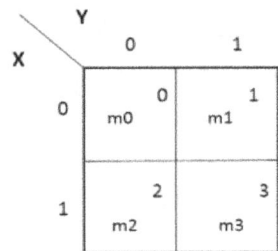

Figure 6.1 : Two Variable K-map

IP#1 Given: F(X,Y) = $\sum_m$ (0,1) = m0 + m1

Solution: From the figure minterms 0 and 1 are grouped into one group corresponding to X = 0 or X'. Remember that a minterm's normal value is 1. Thus, if it assumes a value of 0, it must be complemented. Since in this illustration X =0, it must be X'. Note that X=0 is the same for m0 and m1. It is the only variable that did not change in value for mo and m1. Y changed in value from 0 to 1.

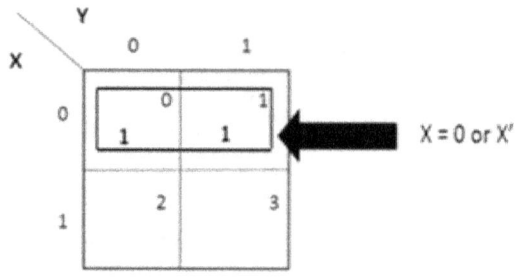

Answer : F = X'

IP#2 Given: F(X,Y) = $\sum_m$ (0,3) = m0 + m3

Solution: From the figure it can be seen that m0 and m3 are not adjacent so each one must be grouped separately.

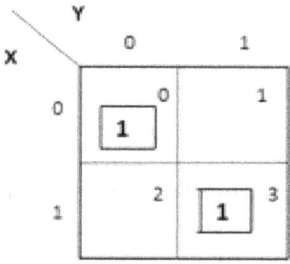

Answer : F(X,Y) = X'Y' + XY

The given is in its simplest form already.

3-VARIABLE K-MAPS

For Boolean expressions of three variables there are eight (2^3) product terms or minterms corresponding to eight squares in the Karnaugh map as shown in Figure 6.3. A minterm present in the expression is represented as 1 in the map. The procedure for k-mapping is similar for two-variables. The grouping starts with the one with the largest number of minterms to the smallest one. The number of variables is determined by the number of minterms in a group as follows :

8 Minterms = "1"

4 Minterms = 1 Variable

2 Minterms = 2 Variables

1 Minterm = 3 Variables

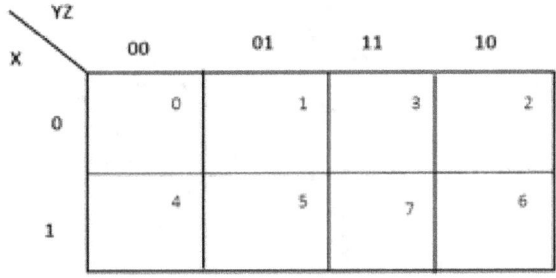

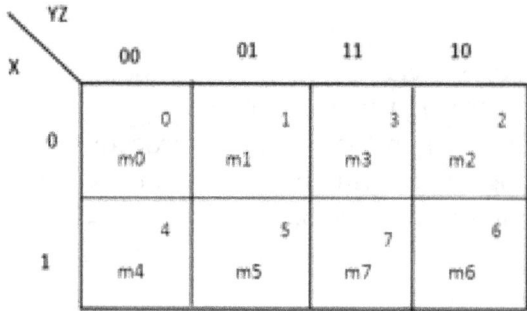

Figure 6.3 Three-Variable Karnaugh Map

IP # 3 Given : **FOUR CORNERS PROBLEM** : $F(X,Y,Z) = \sum_m (0,4,5,7)$

Solution : The four corners are adjacent to each other. Horizontal examination of the minterms show that mo and m4 differs only by one variable, from 000 to 010. The same is true for minterms 5 and 7. From 100 to 110. The variable that did not change is the reduced term which is Z =0.

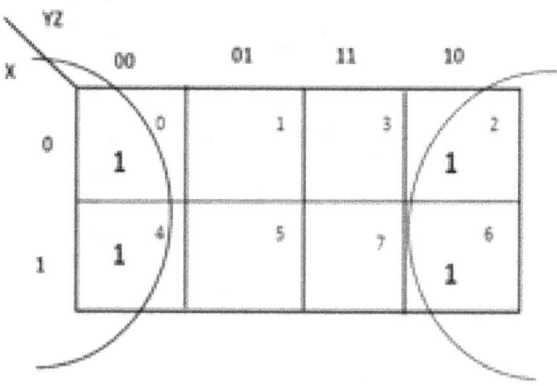

Answer : F(X,Y,Z) = Z'

IP# 4 Given : F (X,Y,Z) = $\sum_m$ (0, 1,6, 7,8)

Solution : A close examination of the minterms in the map yield that there are two groups. Each group consists of two adjacent minterms with only one variable changing. Minterms mo and m1 are adjacent to each other because only one variable Z , changed in value. X and Y both remained at 0. The reduced term form this group is X'Y' . The third group is made up of m5 and m7 where only the variable X changing in value. The reduced term for this group is XZ.

Answer : F(X,Y,Z) = X'Y' + XZ

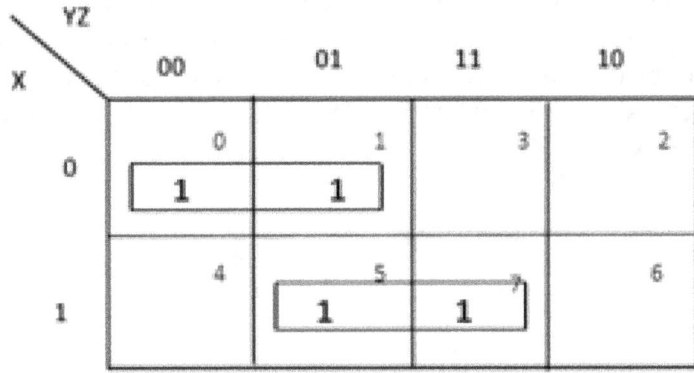

IP # 5 : Given : F(X,Y,Z) = $\sum_m$ (1, 2, 3, 4, 6, 7)

Solution : The largest group is composed of m1, m3 , m5 and m7. In the vertical direction, minterms m1 is adjacent to m5 because in 001 and 101 only X changes in value. Likewise, minterms m3 and m7 are adjacent to each other because in 011 and 111, only X changes in value. In the horizontal direction, m1 and m3 are adjacent to each other in 001 and 011, only Y changes in value. Likewise m5 and m7 are adjacent to each other in 101 and 111, only Y changes in value. The reduced term for the group of four is made up Z . Only minterms m2 and m4 are not yet grouped. Each one can be grouped into two. Minterm m2 can be grouped with m3. The reduced term is XY. Minterm m4 can be group with m5. The reduced term is X'Y. Minterm m4 can be grouped with m5. The reduced term is XY'. There is a tendency to regroup minterms which were grouped already in a bigger group. For example, grouping m1 and m3 or m1 and m5 into groups of two. Note that minterms may be grouped again only if they will make a group bigger like regrouping m5 with m4 or m3 with m2. The simplest form for this problem is F = Z + X'Y' + XY'.

Answer : F = Z + X'Y' + XY'

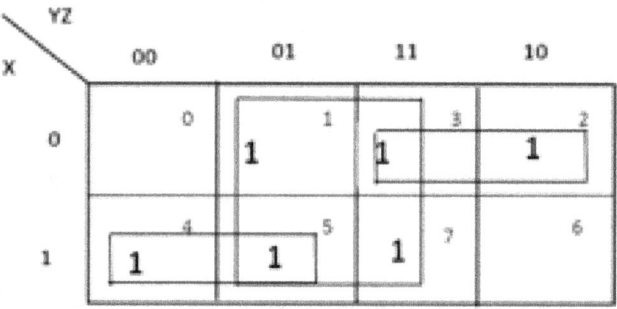

4-VARIABLES K-MAP

For Boolean expressions of four variables there are sixteen (2^4) product terms or minterms corresponding to sixteen squares in the Karnaugh map as shown in Figure_. A minterm present in the expression is represented as 1 in the map. The procedure for k-mapping is similar for three variables. The grouping starts with the one with the largest number of minterms to the smallest one. The number of variables is determined by the number of minterms in a group as follows :

16 Minterms = "1"

8 Minterms = 1 Variable

4 Minterms = 2 Variables

2 Minterms = 3 Variables

1 Minterm = 4 Variables

WX \ YZ	00	01	11	10
00	0	1	3	2
01	4	5	7	6
11	12	13	15	14
10	8	9	11	10

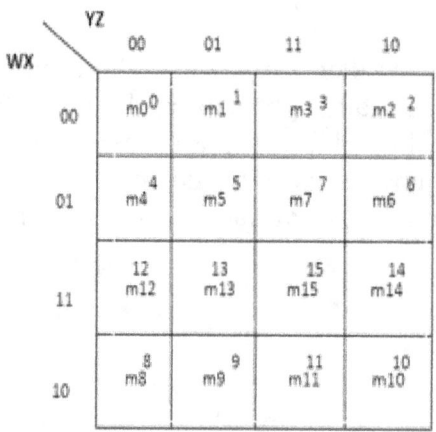

Figure 6.4 Four Variable K-map

IP # 6 : Given : F(W,X,Y,Z) = $\sum_m$ (0,1,4,5,6,7, 8,9,11,12,13)

Solution: An examination of the map shows that there is a group of eight, a group of four and a group of two. The reduced form for the group of eight is Y', the group of four is W'X and the group of two is WX'Y. Note that m4 and m5 are grouped again to make a bigger group of four with m6 and m7. Likewise, m9 is regrouped with m11 to form a group of two.

Answer : F = Y' + W'X + WX'Y

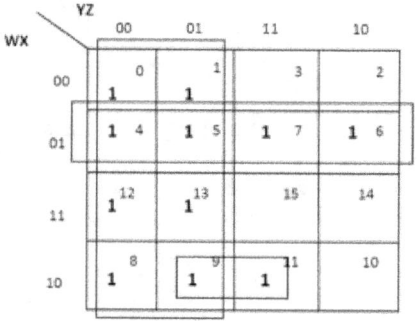

IP#7 :   Given : F(W,X,Y,Z) =∑m(1,3,4,6,9,11)

Solution: The map has two groups of four. The first one is composed of minterms m1, m3, m9 and m11. Minterms m1 and m3 are adjacent to m9 and m11 . The reduced term is X'Y. Likewise, minterms m4 and m12 are adjacent to m6 and m14. The reduced term is XZ'

Answer : F = X'Y + XZ'

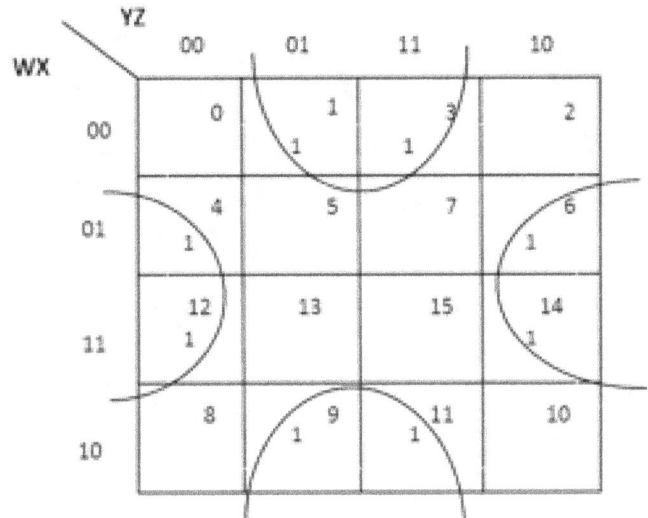

IP # 8 : Given : F(W,X,Y,Z) = ∑m(2,5,7,8,13,15)

Solution : There is a group of four consisting of minterms m5, m7 , m13 and m15. The reduced term is XY. There are two groups of 1. The first one is m2, W'X'YZ' and the second one m8, WX'Y'Z'. Note that minterms belonging to the diagonal are not adjacent to each other. In this illustration, m2, m7, m8 and m13 are not adjacent to each other.

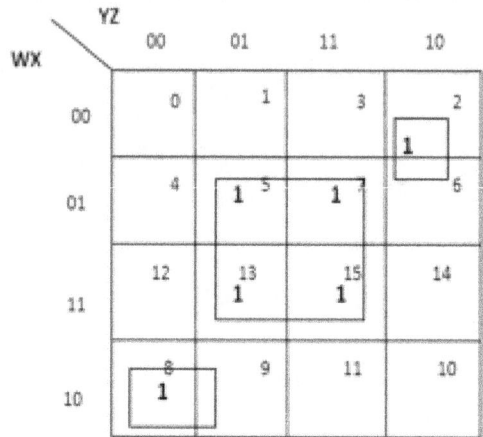

Answer : XY + W'X'YZ' + WX'Y'Z'

Simplifying Product of Sums

The procedure for Product of Sums is the same except that instead of placing a 1 for each term , we placed a zero for each Maxterm on the K-map. The true value of a Maxterm is 0 so a Maxterm assuming a value of 1 must be complemented.

IllustrativeProblem # 9 : F(W,X,Y,Z) =Π_M (2,5,7,8,13,15)

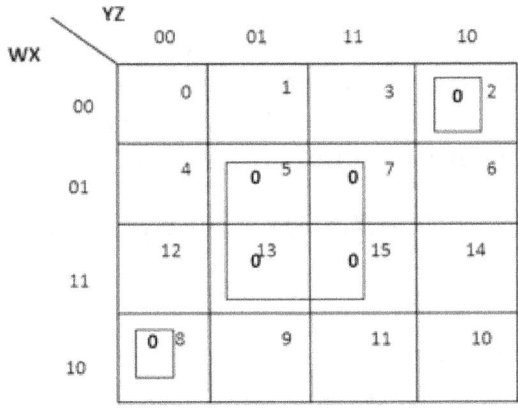

Solution :

There is a group of four consisting of maxterms, M5, M7, M13 and M15. The reduced term is XZ. There are two other groups of one. The first one is maxterm, M2 equivalent to W'X'YZ' and maxterm, M8 equivalent to WX'Y'Z'.

Answer : XZ + W'X'YZ' + WX'Y'Z'

Illustrative Problem # 10 : Given : F(W,X,Y,Z) =Π_M(1,3,4,6,9,11)

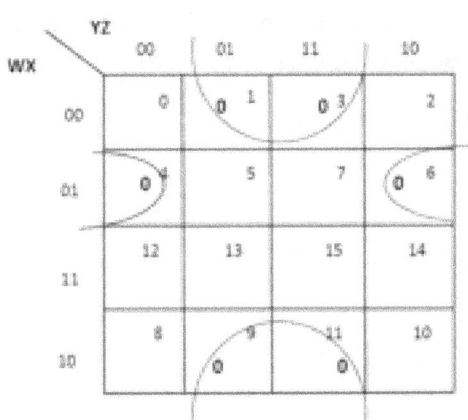

Solution :

There is a group of four consisting of maxterms, M1, M3, M9 and M11. The reduced term is X'Z. The other is a group of two consisting of maxterms M4 and M6. The reduced term is W'XZ'.

Answer : X'Z + W'XZ'

Don't Cares

Don't cares may assume a value of either 1's or 0's. For the purpose of Karnaugh Mapping don't cares are grouped only if they will maximize the size of the grouping. For POS, don't cares are considered to be equal to zeroes. For SOP, don't cares assumed a value of one's, thus it may be grouped with the other minterms. We will use d as the symbol for don't cares.

Illustrative Problem # 11: Given : $F(W,X,Y,Z) = \sum_m(2,5,7,8,13,15), d(3,8)$

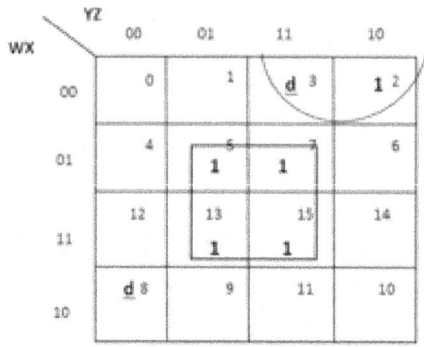

Solution : There is a group of four minterms consisting of m5, m7, m13 and m15. The reduced term is XZ. There is another group of two consisting of a don't care at m3 and minterm, m2. The reduced term is W'X'Y. The don't care at m8 will not be grouped. Recall that don't cares are grouped with minterms only if they will make the grouping bigger and thereby reducing the number of variables in the term.

Answer : F = XZ + W'X'Y

Illustrative Problem # 12 : Given : F(W,X,Y,Z) =$\sum_m$(1,3,4,6,9,11), d (5,7,13,15)

Solution : There is a group of eight consisting of minterms m1, m3, m9 and m11 and don't cares at m5, m7, m13 and m15. The reduced term is Z. The other group is a group of four consisting of m4 and m6 and don't cares at m5 and m7. The reduced term is W'X.

Answer : F = Z + W'X

Illustrative Problem # 13 : Given : F(W,X,Y,Z) =Π_M(1,3,4,6,9,11), d(5,7)

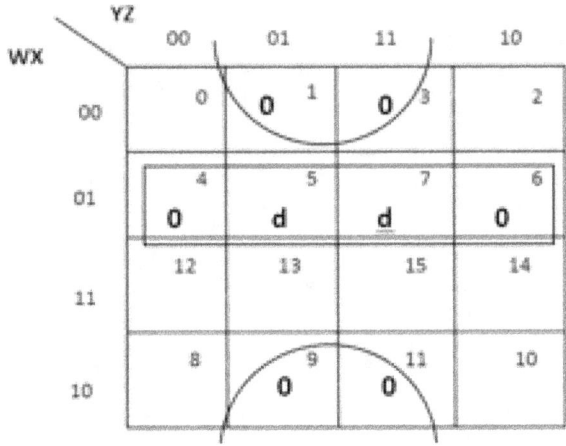

Solution : There are two groups of four. The first one is made up of m1, m3, m9 and m11. The reduced term is X'Z. The other one consists of W'X.

Answer : F = X'Z + W'X

IllustrativeProblem # 14 : $F(W,X,Y,Z) = \Pi_M (2,5,7,8,13,15), d(0,1,14)$

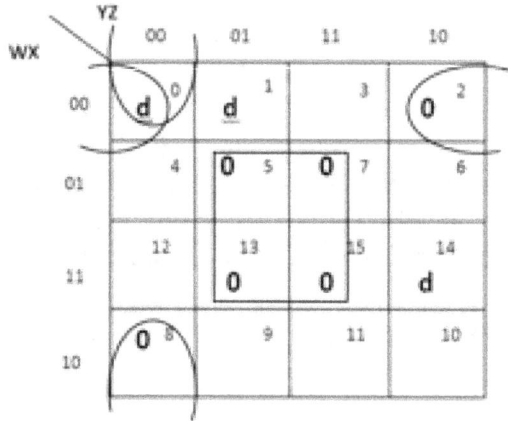

Solution : There is a group of four consisting of minterms m5, m7, m13 and m15. The reduced term is XZ. There are two groups of two. The first one consists of m0 and m8. The reduced term is X'Y'Z'. The second one consists of m0 and m2. The reduced term is W'X'Z'

Answer : F = XZ + X'Y'Z' + W'X'Z'

CHAPTER 6 END OF THE CHAPTER ACTIVITIES :

A. Simplify using K-maps the following Boolean expressions :

1. $F(W,X,Y,Z) = \Pi_M(3,5,7,8,9,15)$

2. $F(W,X,Y,Z) = \Pi_M(1,2,3,4,5,7,11), d(14,15)$

3. $F(X,Y,Z) = \sum_m(1,5,7), d(2,4,6)$

4. $F(X,Y,Z) = \sum_m(2,4,6), d(0,1)$

5. $F(X,Y,Z) = \Pi_M(1,5,7), d(2,4,6)$

6. $F(X,Y,Z) = \Pi_M(2,4,6), d(2,4,6)$

7. $F(W,X,Y,Z) = \sum_m(3,8,9,11), d(0,15)$

8. $F(W,X,Y,Z) = \sum_m(1,13,14,15), d(0,2,4,6)$

9. $F(W,X,Y,Z) = \sum_m(5,7,8,9), d(0,1,2,3)$

10. $F(W,X,Y,Z) = \sum_m(0,5,9), d(1,2,3)$

B. Simplify using K-maps the complements of numbers 1 to 10 of the Boolean functions above.

LOOK UP – Online Activities :

Look up for the Quine-Mckluskey Method and use the same to simplify functions

Look up for more functions to simplify

STEP UP FOR YOUR APP

Make an APP that will diagram on a K-map a Boolean function

Make an APP that will generate the simplified function

EXPERIMENT 5 : MINIMIZATION USING KARNAUGH MAPS

Learning Outcomes :

1. To illustrate and prove using actual circuits that a Boolean function minimized with K-maps is equivalent to the minimized function.

2. To diagram on the appropriate K-map the given Boolean function.

Materials:

Digital Trainer or Breadboard, appropriate power supply (+5V) , LEDs to monitor the output. Integrated Circuits (AND gate – 7408), (INVERTER – 7404), (OR gate – 7432), connecting wires. Data Sheets.

Procedure :

Step 1: Given the Boolean function:

$F(X,Y,Z) = \sum_m (1, 2, 3, 4, 6, 7) = X'Y'Z' + X'YZ' + X'YZ + XY'Z' + XYZ' + XYZ$

1.a Draw and wire the Circuit.

1.b Generate the Truth table based on actual output measured from the experimental circuit.

Step 2 : Minimize the given function in Step 1 using Karnaugh maps.

Ans. _____

2.a Draw and Wire the Circuit

2.b Generate the Truth Table based on the actual output measured from the experimental circuit.

Step 3 : Compare the truth table diagrammed in Step 1.b and Step 2.b. Are they the same ? Why ?

_____
_____
_____

CONCLUSION

_____
_____
_____
_____

CHAPTER 7 - COMBINATIONAL LOGIC CIRCUIT ANALYSIS

Learning Outcomes :

1. To learn the step by step procedure for analyzing a Combinational Logic Circuit (CLC)

2. To be familiar with the analysis.

3. To verify the correctness of the analysis.

4. To infer that analysis and design of CLC are interrelated and that they are a cyclical process.

GENERAL ANALYSIS PROCEDUR : STEP BY STEP

For the analysis of CLC, the logic diagram is given. From the diagram, analysis proceeds as follow.

Step 1 : From the given diagram, the output equation is derived by moving through each intermediate point until the output is reached.

    1.1 Label each input. The number of inputs, n, determine the number of all possible Combinations, N. $N = 2^n$

    1.2 Label each Intermediate Point.

    1.3 Generate the equation point by point until the output is reached.

Step 2. A truth table is then generated based on the number of inputs and the output equation. The number of inputs determine the number of all possible combinations, 2^n where n corresponds to the number of inputs.

Step 3. Derive the output equation from the truth table in abbreviated SOP Form.

The circuit is said to be fully analyzed when the truth table is completed and the output expression derived. The table shows the given output for all possible input combinations.

An OBE Approach to Logic Circuits and Digital Design: Step by Step

Illustrative Problem # 1: FULL ADDER (EXPANDED) CIRCUIT ANALYSIS

The full adder may be implemented in its simplest form using Exclusive OR Gates. This example uses a combination of AND, INVERTER and OR gates only.

Given : Expanded Full Adder Diagram

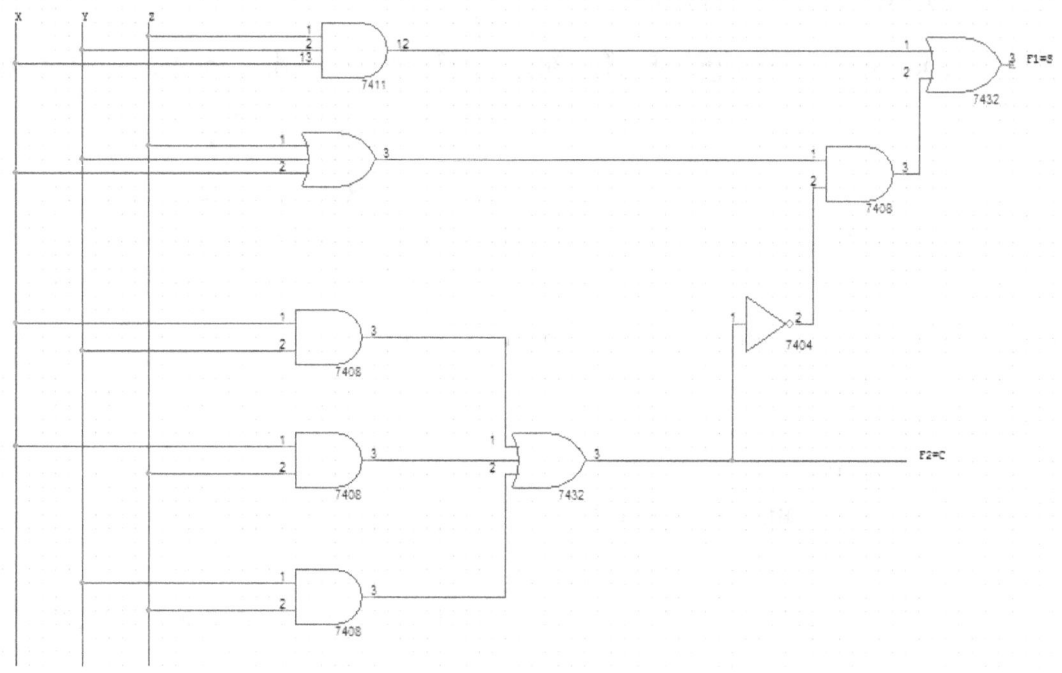

Step 1 : From the given diagram, the output equation is derived by moving through each intermediate point until the output is reached.

At Point P1:

P1 (X,Y,Z) = XY + XZ + YZ

At Point P2 :

P2(X,Y,Z) = (X+Y+Z)P1'

P2(X,Y,Z) = (X+Y+Z)P1'

At F1 = S, the Sum Output

S = (XYZ) P2

At F2 = C, the Carry Output

F2=C=P1=XY + XZ + YZ

Step 2 : Generate Truth Table including Intermediate Points

index	X	Y	Z	P1	P2	F1=S	F2=C
0	0	0	0	0	0	0	0
1	0	0	1	0	1	1	0
2	0	1	0	0	1	1	0
3	0	1	1	1	0	0	1
4	1	0	0	0	1	1	0
5	1	0	1	1	0	0	1
6	1	1	0	1	0	0	1
7	1	1	1	1	0	1	1

Step 3: Simplified output Expression in SOP:

$C(X,Y,Z) = \sum_m(3,5,6,7)$

$S(X,Y,Z) = \sum_m(1,2,4,7)$

An OBE Approach to Logic Circuits and Digital Design: Step by Step

Illustrative Problem # 2 : FULL-SUBTRACTOR ANALYSIS

Given : Full Subtractor Diagram

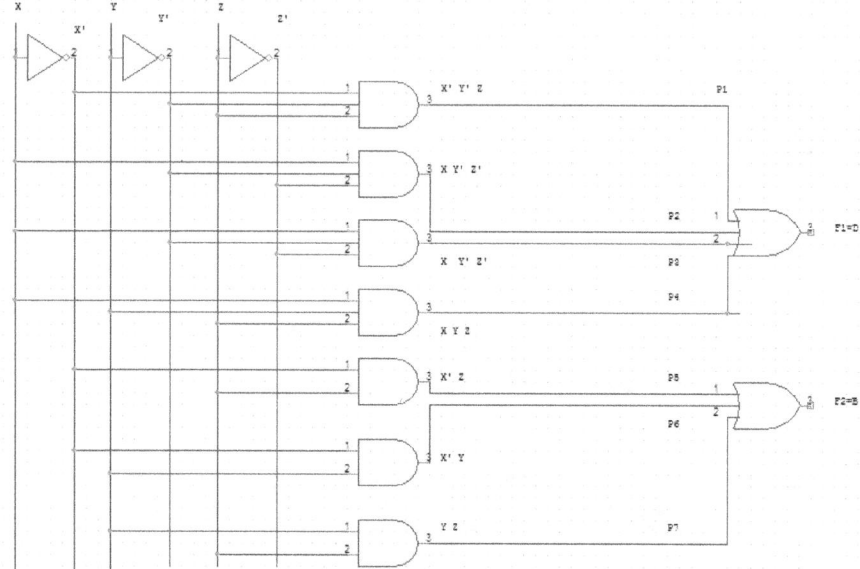

Step 1 : From the given diagram, the output equation is derived by moving through each intermediate point until the output is reached.

At points P1 to P4 to generate output F1 = D, the Difference output for the Subtractor.

$P1 = X'Y'Z$

$P2 = X'YZ'$

$P3 = XY'Z'$

$P4 = XYZ$

$F1 = D(X,Y,Z) = P1 + P2 + P3 + P4$

At points P5 to P7 to generate output F2 = B, the Borrow output for the Subtractor.

$P5 = X'Z$

$P6 = X'Y$

$P7 = YZ$

$F2 = B(X,Y,Z) = P5+P6+P7$

An OBE Approach to Logic Circuits and Digital Design: Step by Step

Step 2. A truth table is then generated based on the number of inputs and the output equation.

index	X	Y	Z	P1=X'Y'Z	P2=X'YZ'	P3=XY'Z'	P4=XYZ	P5=X'Z	P6=X'Y	P7=YZ	F1	F2
0	0	0	0	0	0	0	0	0	0	0	0	0
1	0	0	1	1	0	0	0	1	0	0	1	1
2	0	1	0	0	1	0	0	0	1	0	1	1
3	0	1	1	0	0	0	0	1	1	1	0	1
4	1	0	0	0	0	1	0	0	0	0	1	0
5	1	0	1	0	0	0	0	0	0	0	0	0
6	1	1	0	0	0	0	0	0	0	0	0	0
7	1	1	1	1	0	0	1	0	0	1	1	1

Step 3 : Abbreviated output Expression in SOP from the truth table

F1 = D (X,Y,Z) = $\sum$m (1,2,4,5,7)

F2= B (X,Y,Z) = $\sum$m (1,2,3,7)

The analysis is done with the final SOP Equation describing the circuit and the complete truth table describing the output for each possible input combination

Illustrative Problem # 3 : Analyze the Logic Diagram Below:

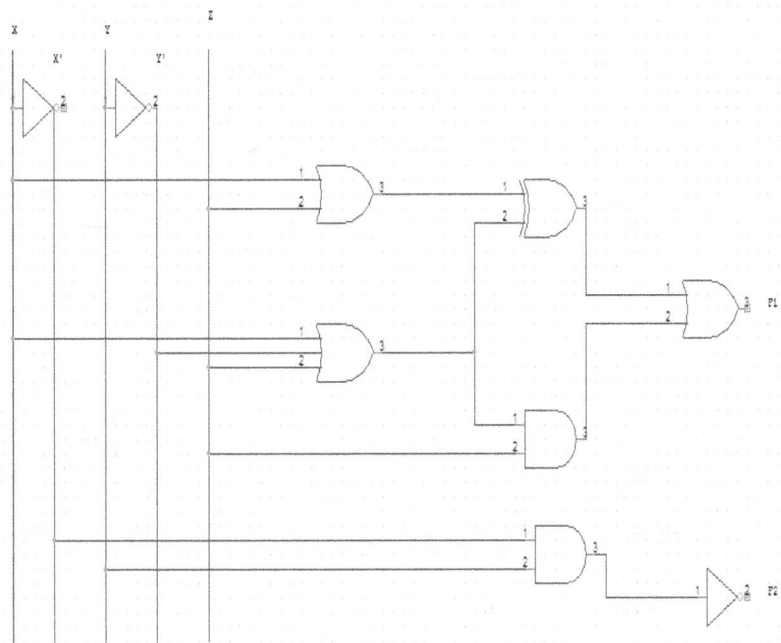

Step 1 :From the given diagram , the output equation is derived by moving through each intermediate point until the output is reached.

At points P1, P2, P3 and P4 to generate F1

P1 = XZ

P2 = XY'Z

P3 = P1 xor P2

P4 = (P2)(Z)

F1 (X,Y,Z) = P3 + P4

At point P5 to generate F2

P5 = X'Y

F2 = P5' = (X'Y)'

Step 2 : Generate the Truth Table including the Intermediate Points.

index	X	Y	Z	P1=XZ	P2=XY'Z	P3=P1xorP2	P4=(P2)(Z)	P5=X'Y	F1=P3+P4	F2=P5'
0	0	0	0	0	0	0	0	0	0	1
1	0	0	1	0	0	0	0	0	0	1
2	0	1	0	0	1	1	0	1	1	0
3	0	1	1	0	0	0	0	1	0	0
4	1	0	0	0	0	0	0	0	0	1
5	1	0	1	1	0	1	0	0	1	1
6	1	1	0	0	0	0	0	0	0	1
7	1	1	1	1	0	1	0	0	1	1

Step 3: Abbreviated output Expression in SOP :

F1(X,Y,Z) = $\sum$m (2,5,7)

F2(X,Y,Z) = $\sum$m (0,1,4,5,6,7)

CHAPTER 7 END OF THE CHAPTER ACTIVITIES:

ANALYZE THE FOLLOWING CIRCUITS

1.

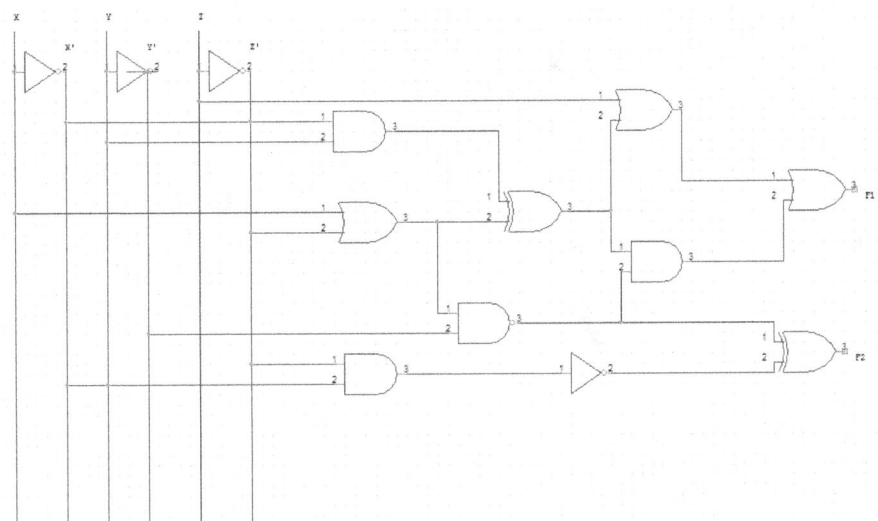

2.

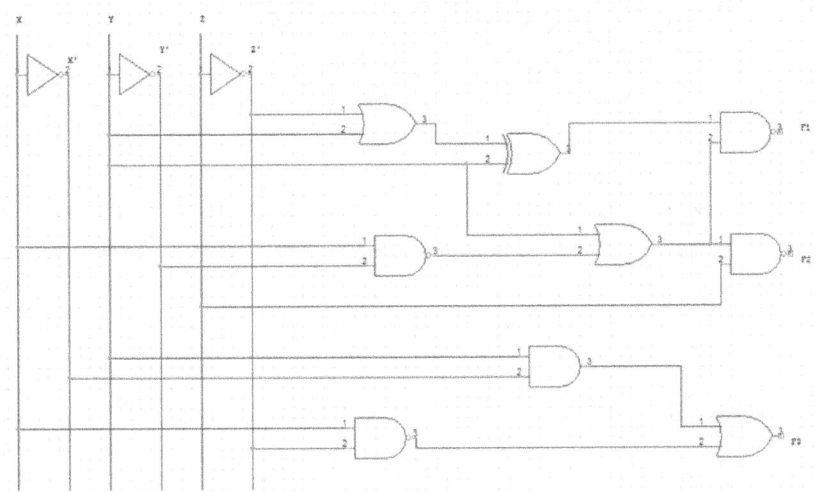

3.

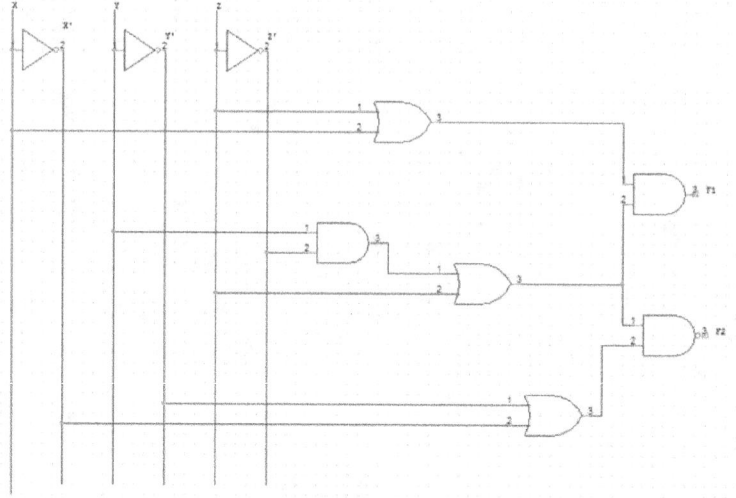

4.

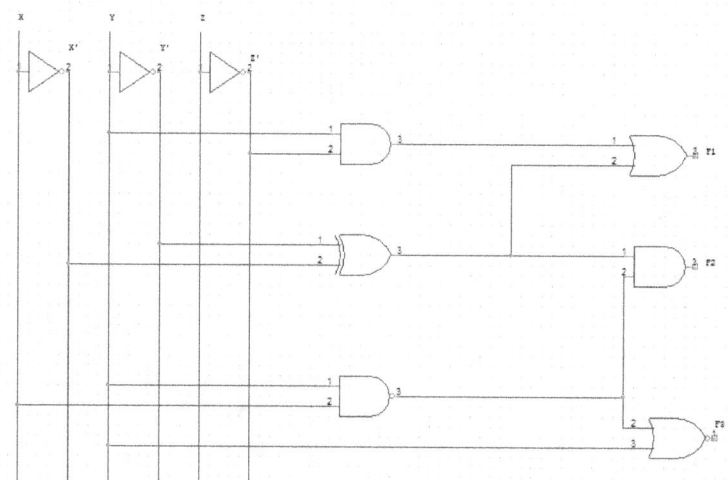

5.

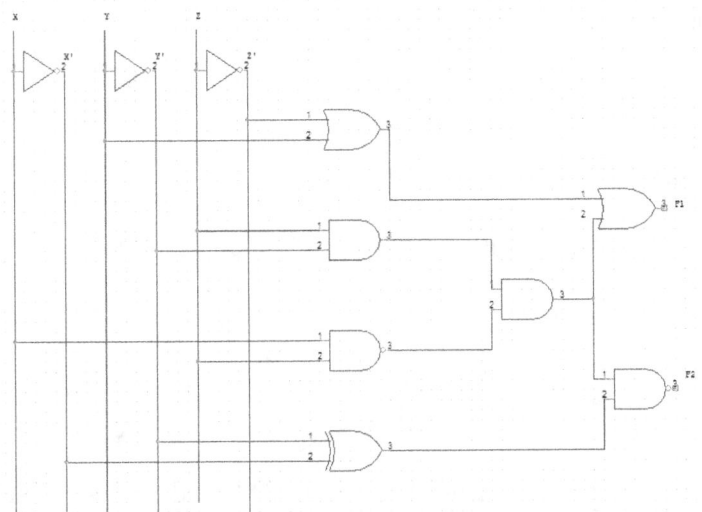

EXPERIMENT 6 – ANALYSIS OF A COMBINATIONAL LOGIC CIRCUIT (CLC) - FULL SUBTRACTOR CLC ANALYSIS

Learning Outcomes :

1. To analyze a Full Subtractor circuit

2. To construct a Full Subtractor Circuit.

3. To identify each Intermediate Point and observe the actual state of each LED.

Materials :

Digital Trainer or Breadboard, appropriate power supply (+5V), LEDs and series resistors to monitor the output. Integrated Circuits (AND gate – 7408), (INVERTER – 7404), (OR gate – 7432), connecting wires. Data Sheets.

Procedure :

Step 1 : Wire the Experimental Circuit Below :

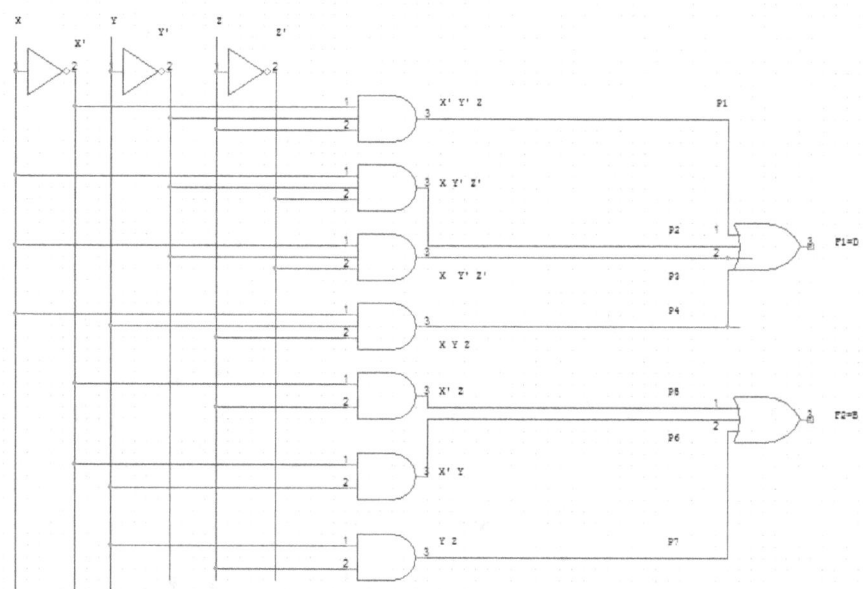

Step 2: Generate the Truth Table by recording the state of the LEDs for each of the sixteen possible combinations for the four inputs W, X, Y and Z.

Step 3 : Compare the Actual Truth table with the Theoretical Truth Table. Are they the same? Why or Why not ?

_____
_____
_____
_____

CONCLUSION:

_____
_____
_____

CHAPTER 8 - COMBINATIONAL LOGIC CIRCUIT DESIGN

Learning Outcomes :

1. To describe the step by step procedure of designing a combinational logic circuit

2. To identify the tools required in the design process and use them effectively and efficiently.

3. To explain the link between analysis and design of combinational logic circuit

4. To explain the operation of standard CLCs .

The design of a combinational logic circuit starts with the specifications and ends up with the circuit. The specifications may come in the form of a worded description of the circuit's function , a truth table or a Boolean expression.

DESIGN PROCEDURE : STEP BY STEP

1. Determine from the specifications the number of inputs.

2. Generate the truth table.

3. Determine the simplified expression from the truth table in either Sum of Products (SOP) or Product of Sums (POS) abbreviated form.

4. Simplify using K-map

5. Draw the logic Diagram

6. Verify the correctness of the design

The design process will be illustrated using the following standard circuits :

1. Code Converters

2. Adders and Subtractors

3. Decoders

4. Encoders

5. Multiplexers

6. Demultiplexers

CODE CONVERTERS

Code converters are CLCs that convert one binary code to another. Examples are Binary Coded Decimal (BCD) to Gray Code Converter, BCD to Excess-3 and BCD to 7 segment decoder among others.

Illustrative Problem # 1 : Design a logic circuit that will convert the decimal numbers 0 to 15 to Gray Code

Gray code finds application in shaft converters and analog to digital converters. It belongs to a class of code known as minimum change code. The code is formed by retaining the most significant bit of the binary number being converted and taking the exclusive-or of the succeeding bits. One will notice that only one bit changes for each number in succession.

Step 1 : Since the decimal numbers 0 to 15 which has 16 numbers will be converted, four (4) input bits will be required. Recall that $2^n = N$ where n refers to the number of inputs and N the number of all possible combinations.

An OBE Approach to Logic Circuits and Digital Design: Step by Step

Step 2. Generate the truth table.

DECIMAL	BINARY				GRAY			
	W	X	Y	Z	A	B	C	D
0	0	0	0	0	0	0	0	0
1	0	0	0	1	0	0	0	1
2	0	0	1	0	0	0	1	1
3	0	0	1	1	0	0	1	0
4	0	1	0	0	0	1	1	0
5	0	1	0	1	0	1	1	1
6	0	1	1	0	0	1	0	1
7	0	1	1	1	0	1	0	0
8	1	0	0	0	1	1	0	0
9	1	0	0	1	1	1	0	1
10	1	0	1	0	1	1	1	1
11	1	0	1	1	1	1	1	0
12	1	1	0	0	1	0	1	0
13	1	1	0	1	1	0	1	1
14	1	1	1	0	1	0	0	1
15	1	1	1	1	1	0	0	0

Step 3 : Determine the abbreviated expression from the truth table in Sum of Products (SOP) form.

$A(W,X,Y,Z) = \sum_m (8,9,10,11,12,13,14,15)$

$B(W,X,Y,Z) = \sum_m (4,5,6,7,8,9,10\ 11)$

$C(W,X,Y,Z) = \sum_m (2,3,4,5,10,11,12,13)$

$D(W,X,Y,Z) = \sum_m (1,2,5,6,9,10,13,14)$

Step 4 : Simplify using Karnaugh maps

The simplified functions are the following:

A = W

B = W'X + WX' = W xor X

C = XY' + X'Y = X xor Y

D = Y'Z + YZ' = Y xor Z

Draw the Logic Diagram

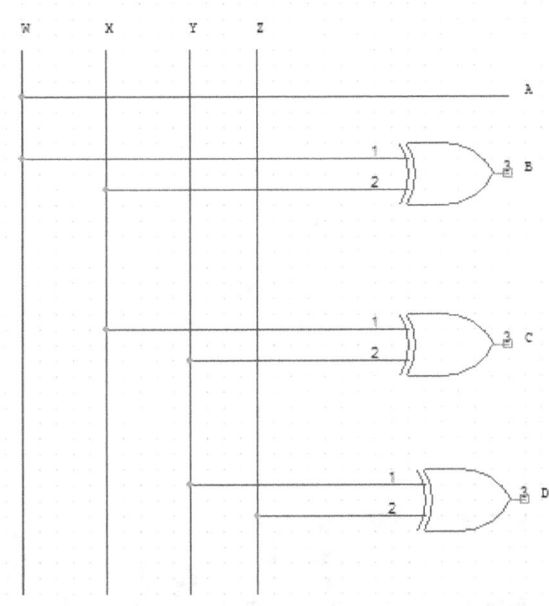

6. Verify the correctness of the design

Step 6.1 From the table, test values :

#6 W=0, X=1, Y=1, Z=0

Step 6.2 Apply on the logic diagram in Step 5.

Step 6.3 Get the output

A=0, B=1, C=0, D=1

Note that the Circuit output matches the truth table entry.

The seven segment decoder finds application in digital readouts for watches, calculators, appliances using LEDs. This code converter accepts a digital digit in Binary Coded Decimal (BCD) format and generates the appropriate outputs for the selection of the segments.

Illustrative Problem # 2 : BCD - to - SEVEN Segment Decoder

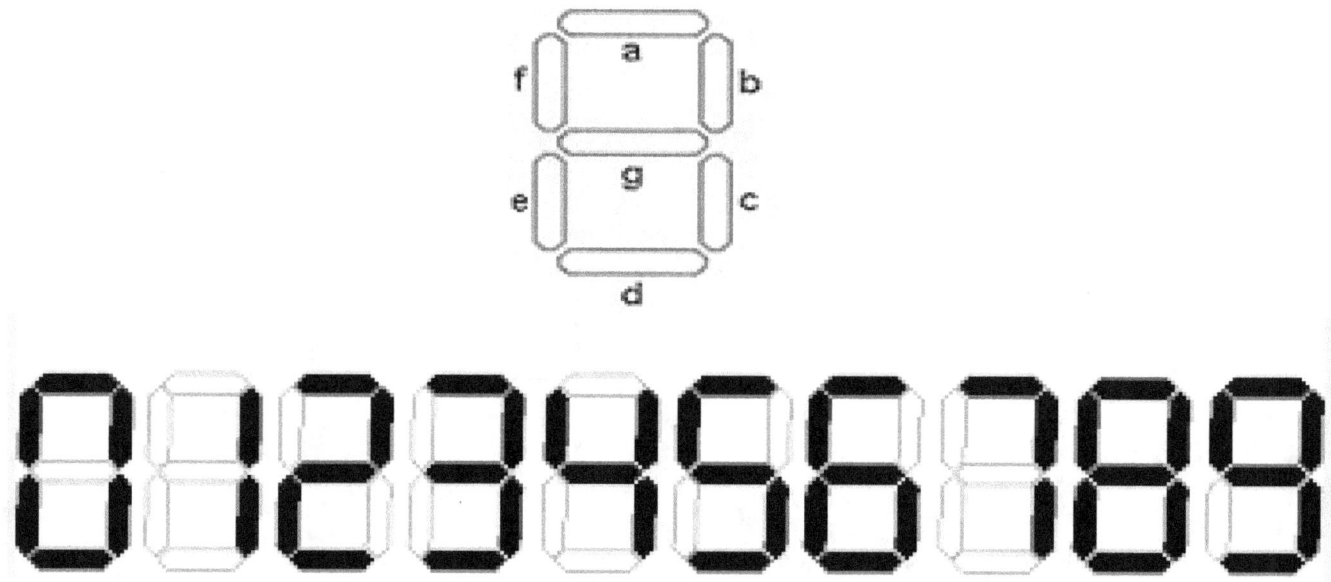

Requirements: Design a BCD-to-seven segment decoder . Decimal numbers 0 to 9 are converted to their seven segment equivalents. All other inputs beyond 9 are equal to zero.

Step 1. Determine from the specifications the number of inputs.

$$2^n >= N$$

0 to 9 means 10 combinations, N= 10

2^4 must be greater than or equal to 10

n = 4; there are four (4) input bits

Step 2 : Generate the truth table

DECIMAL	BCD				a	b	c	d	e	f	g
	W	X	Y	Z							
0	0	0	0	0	1	1	1	1	1	1	0
1	0	0	0	1	0	1	1	0	0	0	0
2	0	0	1	0	1	1	0	1	1	0	1
3	0	0	1	1	1	1	1	1	0	0	1
4	0	1	0	0	0	1	1	0	0	1	1
5	0	1	0	1	1	0	1	1	0	1	1
6	0	1	1	0	1	0	1	1	1	1	1
7	0	1	1	1	1	1	1	0	0	0	0
8	1	0	0	0	1	1	1	1	1	1	1
9	1	0	0	1	1	1	1	0	1	1	1
	All	other	inputs		0	0	0	0	0	0	0

Step 3 : Determine the abbreviated expression from the truth table in Sum of Products (SOP) form.

a = $\sum$m (0,2,3,5,6,7,8,9)

b = $\sum$m (0 to 4, 7 ,8, 9)

c = $\sum$m (0,1, 3 to 9)

$d = \sum m\ (0,2,3,5,6,8)$

$e = \sum m\ (0,2,6,8,9)$

$f = \sum m\ (0,4,5,6,8,9)$

$g = \sum m\ (2,3,4,5,6,8,9)$

Step 4 : Simplify using Karnaugh Map.

The following are the simplified functions:

a = W'Y + W'XY' + W'X'Z' + WX'Y'

b = W'X' + W'Y'Z + W'YZ + WX'Y'

c = W'X + W'Z + W'X'Y + WX'Y'

d = W'X'Y + W'YZ' + X'Y'Z' + W'XY'Z

e = W'YZ' + W'X'Z' + WX'Y'

f = W'XY' + WX'Y' + W'XZ' + W'Y'Z'

g = W'X'Y + W'XY' + W'XZ' + WX'Y'

The following are simpler functions considering 10 to 15 as don't cares:

a = W + YZ + XZ + X'Z'

b = X' + Y'Z' + YZ

c = X + Y' + Z

d = X'Z' + YZ' + X'Y + XY'Z

e = X'Y' + YZ'

f = W + Y'Z' + XY' + XZ'

g = W + XY' + YZ' + X'Z

ADDERS

A half adder adds two binary bits. A full adder adds two binary bits including the carry in bit. The output consists of the Sum of the bits and the Carry.

Illustrative Problem # 3 : Design a Full Adder

Step 1 : The number of input bits is three (3), n=3 . The number of possible combinations for 3 inputs is eight , $2^3 = 8$. Let X be the Carry in bit and Y ,Z the bits to be added.

Step 2 : Generate the Truth Table

index	X	Y	Z	S	C
0	0	0	0	0	0
1	0	0	1	1	0
2	0	1	0	1	0
3	0	1	1	0	1
4	1	0	0	1	0
5	1	0	1	0	1
6	1	1	0	0	1
7	1	1	1	1	1

Step 3 : Determine the abbreviated expression from the truth table in Sum of Products (SOP) form.

$F1 = S(X,Y,Z) = \sum m (1,2,4,7)$

$F2 = C(X,Y,Z) = \sum m (3,5,6,7)$

Step 4 : Simplify Boolean expression using K-map and Boolean Identities

$F1 = S(X,Y,Z) = X'Y'Z + X'YZ' + XY'Z' + XYZ$

$\quad\quad\quad = X'(Y'Z + YZ') + X(Y'Z' + YZ)$

$\quad\quad\quad = X'(Y \text{ xor } Z) + X(Y \text{ xor } Z)'$

$\quad\quad\quad = X \text{ xor } Y \text{ xor } Z$

$F2 = C(X,Y,Z) = YZ + XZ + XY$

$\quad\quad\quad = YZ + X(Y + Z)$

$\quad\quad\quad = YZ + X(YZ + YZ' + YZ + Y'Z)$ Recall $Y(Z+Z') = Y$ bec $(Z+Z')=1$

$\quad\quad\quad = YZ + X(YZ + YZ' + Y'Z)$ $YZ + YZ = YZ$ from $x+x = 1$

$\quad\quad\quad = YZ + XYZ + X(YZ' + Y'Z)$

$\quad\quad\quad = YZ + X(Y \text{ xor } Z)$ $YZ(1+X) = YZ$

STEP 5 : Draw the circuit

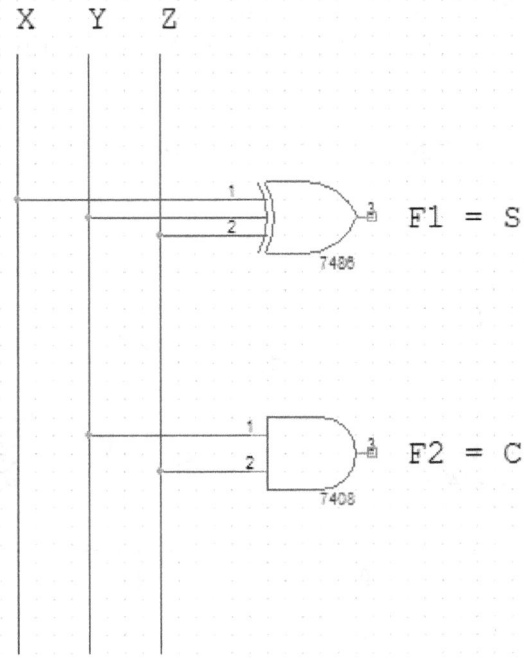

SUBTRACTORS

A half subtractor takes the difference of two binary bits X and Y. The output consists of the Difference, D and the Borrow, B. The full subtractor operates on three input bits corresponding to the minuend, subtrahend and the borrow from the previous subtraction operation. The output consists of the Difference, D and the Borrow, B.

Illustrative Problem # 4 : Design a half subtractor

Step 1 : There are two inputs

Step 2 : Generate the truth table

index	X	Y	D	B
0	0	0	0	0
1	0	1	1	1
2	1	0	1	0
3	1	1	0	0

Step 3 : Determine the abbreviated expression from the truth table in Sum of Products (SOP) form.

$F1 = D(X,Y) = \sum m(1,2)$

$F2 = B(X,Y) = \sum m(1)$

Step 4 : Simplify using Karnaugh Map

The simplified functions are :

$D(X,Y) = X'Y + XY'$

$B(X,Y) = X'Y$

Step 5 : Draw the Logic Diagram

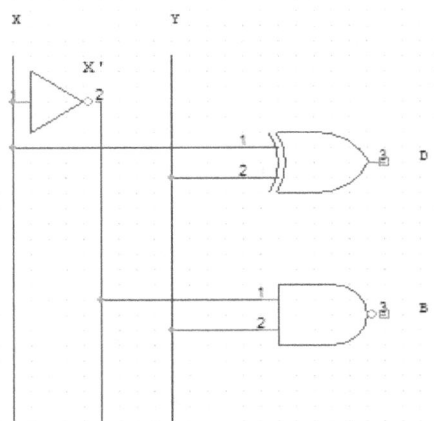

Illustrative Problem # 5 : Design a Full Subtractor

Step 1 : There are three (3) input variables

Step 2 : Generate the Truth Table

index	X	Y	Z	D	B
0	0	0	0	0	0
1	0	0	1	1	1
2	0	1	0	1	1
3	0	1	1	0	1
4	1	0	0	1	0
5	1	0	1	0	0
6	1	1	0	0	0
7	1	1	1	1	1

Step 3 : Determine the abbreviated expression from the truth table in Sum of Products (SOP) form.

F1 = D(X,Y,Z) = $\sum_m(1,2,4,7)$

F2 = B(X,Y,Z) = $\sum_m(1,2,3,7)$

Step 4 : Simplify using Karnaugh Map

The simplified functions are :

D(X,Y,Z) = X xor Y xor Z

B(X,Y,Z) = X'Y + Z (X xor Y)'

Step 5 : Draw the Circuit

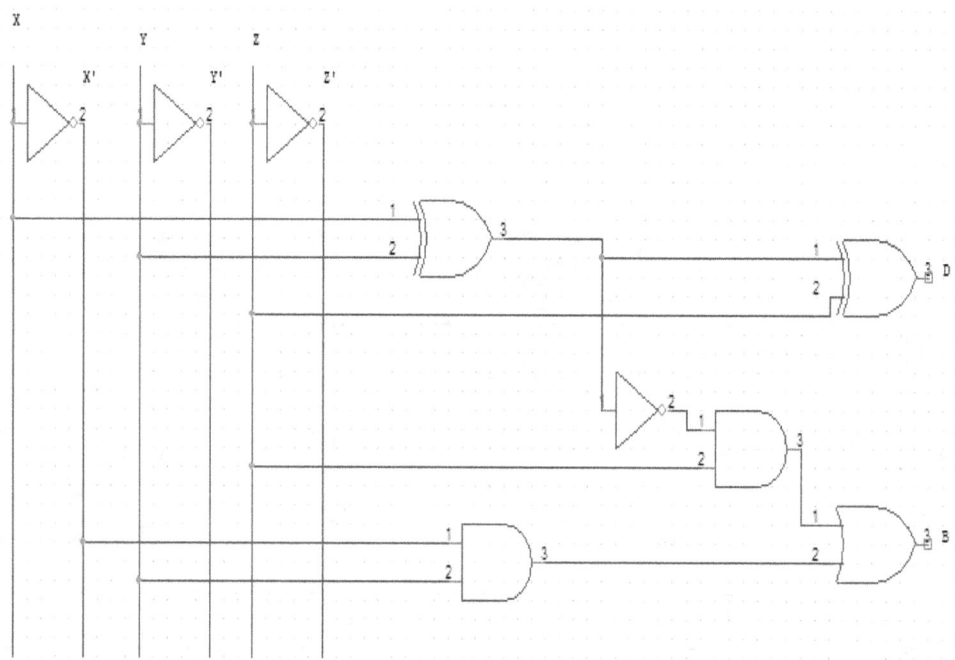

DECODERS

A decoder is a combinational circuit with n input bits and 2^n output lines. It is an n to $N = 2^n$ circuit. Each output corresponds to one and only one output. It is a circuit with few inputs and many outputs. It is a "few-to-many" circuit.

Illustrative Problem # 6 : Design a 3-to-8 Decoder

Step 1 : There are three (3) input variables.

Step 2 : Generate the Truth Table

index	X	Y	Z	D_0	D_1	D_2	D_3	D_4	D_5	D_6	D_7
0	0	0	0	1	0	0	0	0	0	0	0
1	0	0	1	0	1	0	0	0	0	0	0
2	0	1	0	0	0	1	0	0	0	0	0
3	0	1	1	0	0	0	1	0	0	0	0
4	1	0	0	0	0	0	0	1	0	0	0
5	1	0	1	0	0	0	0	0	1	0	0
6	1	1	0	0	0	0	0	0	0	1	0
7	1	1	1	0	0	0	0	0	0	0	1

Steps 3 and 4 . Abbreviated SOP Expression and simplified Boolean expression. No need to simplify using K- maps since each output corresponds to one product term only.

$D0(X,Y,Z) = \sum_m(0) = X'Y'Z'$

$D1(X,Y,Z) = \sum_m(1) = XYZ$

$D2(X,Y,Z) = \sum_m(2) = X'YZ'$

$D3(X,Y,Z) = \sum m(3) = X'YZ$

$D4(X,Y,Z) = \sum m(4) = XY'Z'$

$D5(X,Y,Z) = \sum m(5) = XY'Z$

$D6(X,Y,Z) = \sum m(6) = XYZ'$

$D7(X,Y,Z) = \sum m(7) = XYZ$

Step 5 : Draw the Logic Diiagram

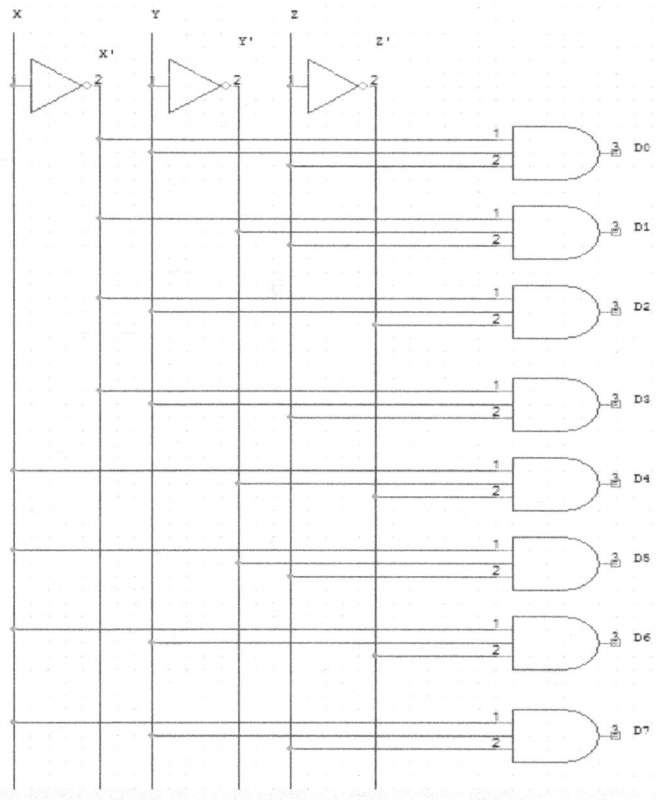

Illustrative Problem # 7 : Design a Full Subtractor using a 3 - to 8 Decoder

Step 1 : Three inputs

Step 2 : From the Truth Table in Illustrative Problem # 5:

Step 3 :

$F1 = D(X,Y,Z) = \sum m(1,2,4,7)$

$F2 = B(X,Y,Z) = \sum m(1,2,3,7)$

The 3-to-8 decoder circuit usually comes in a functional block as shown:

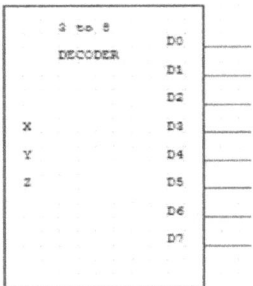

Step 5 : Draw the logic diagram using the 3-to-8 Decoder functional block. A functional block corresponds to one Integrated Circuit (IC). This is level 2 design. In level 1, we use basic gates purely. In level 2, we use functional blocks with basic gates.

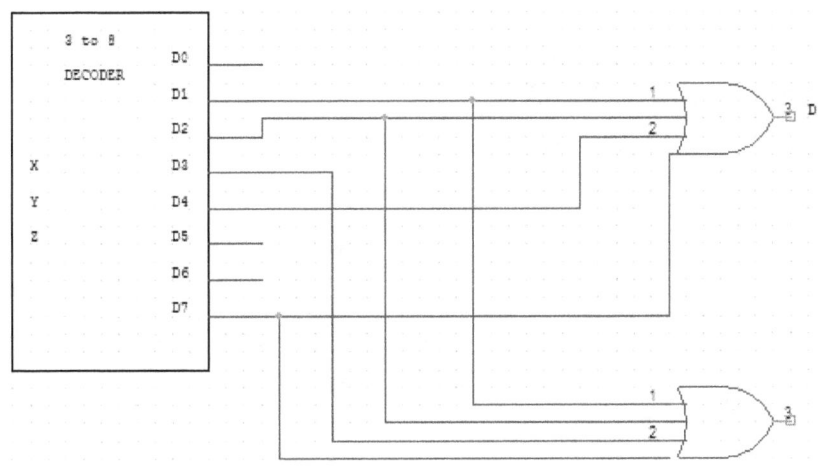

Illustrative Problem #8 : Design a Half-Adder using a 3-to-8 Decoder

Step 1 : There are two inputs, X and Y.

Step 2 : Generate the Truth Table

index	X	Y	D	B
0	0	0	0	0
1	0	1	1	1
2	1	0	1	0
3	1	1	0	0

$F1 = D(X,Y) = \sum_m (1,2)$

$F2 = B(X,Y) = \sum_m (1)$

Skip Step 4

Step 3 : Determine the abbreviated expression from the truth table in Sum of Products (SOP) form.

Step 5 : Draw the logic diagram using a 3-to-8 Decoder

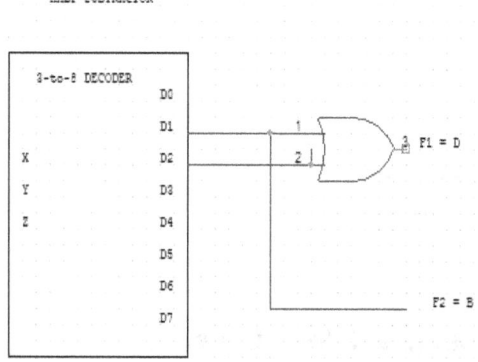

Higher Order Decoders may be constructed from lower order ones. This technique becomes handy when lower order decoders are the only ones available.

Illustrative Problem # 9 : Design a 3-to-8 Decoder from 2-to-4 Decoders

Step 1 : Number of inputs = 3

Step 2 : Determine the number of lower order decoders required.

    A 3-to-8 decoder would require two 2-to-4 decoders.

Step 3 : Draw the logic diagram. Another logic required is how to wire the two lower order decoders. This is done using the Enable Line. When E=0, the 1st 2-to-4 decoder is chosen, when E=1 the second 2-to-4 decoder is chosen.

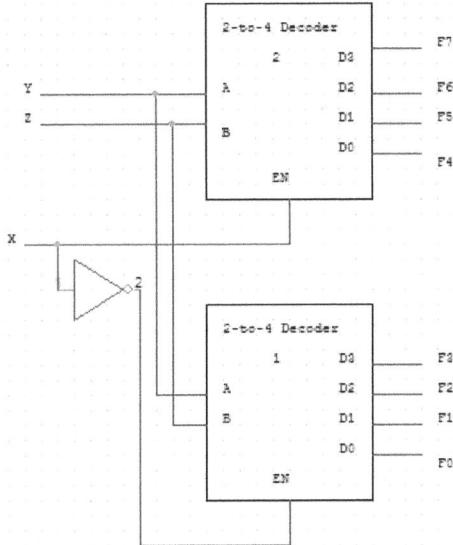

An OBE Approach to Logic Circuits and Digital Design: Step by Step

Illustrative Problem # 10. Design a 4-to-16 Decoder from 3-to-8 Decoders

Step 1 : Number of inputs = 4

Step 2 : Determine the number of lower order decoders required.

A 4-to-16 decoder would require two 3-to-8 decoders

Step 3 : Draw the logic diagram.

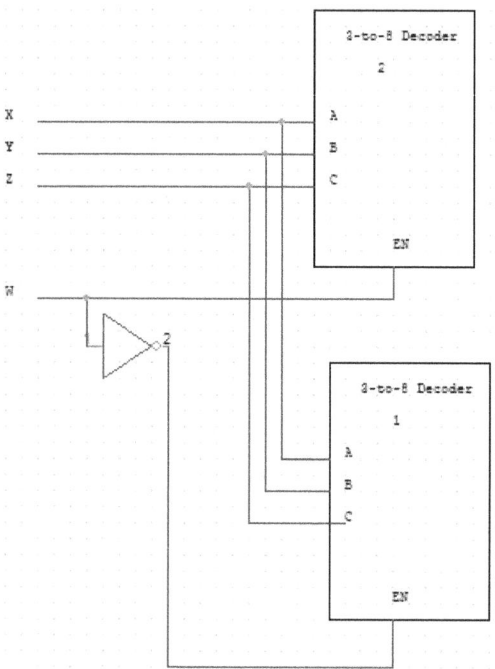

ENCODERS

An encoder performs function that is opposite of the decoder. It is a circuit with many inputs and few outputs. We can call it a many-to-few circuit. The number of inputs, N is equal to 2^n and the number of outputs is equal to n. Note that for each input combination, only one of the outputs go high. The basic encoders are implemented using OR gates and inverters.

Challenges of an Encoder:

When any two inputs go high at the same time, the output will be erroneous. Take for example, D1 and D2 going high at the same time in a 4-to-2 Encoder. Notice that the output is neither 01 nor 10 but 11 which is incorrect. To address this challenge, there must be a way of prioritizing one of the inputs going High at the same time.

PRIORITY ENCODER

The priority encoder assigns a priority to each input such that the one with the highest priority is chosen at any time there is a conflict. For example, in the implementation that follows, the input at the most significant bit is given the highest priority, such that if D6, D5, and D3 go high at the same time, D6 which has the highest priority will be processed first.

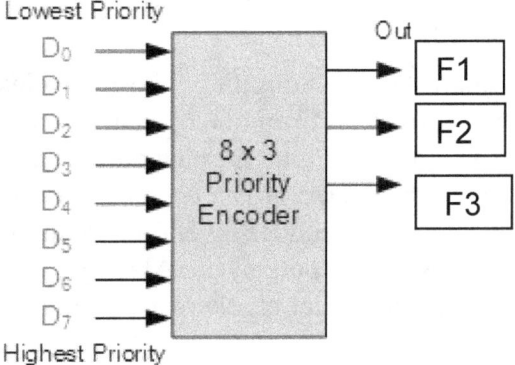

D$_7$	D$_6$	D$_5$	D$_4$	D$_3$	D$_2$	D$_1$	D$_0$	F1	F2	F3
0	0	0	0	0	0	0	1	0	0	0
0	0	0	0	0	0	1	x	0	0	1
0	0	0	0	0	1	x	x	0	1	0
0	0	0	0	1	x	x	x	0	1	1
0	0	0	1	x	x	x	x	1	0	0
0	0	1	x	x	x	x	x	1	0	1
0	1	x	x	x	x	x	x	1	1	0
1	x	x	x	x	x	x	x	1	1	1

MULTIPLEXERS

A multiplexer is a "many-to-one" circuit". The multiplexer is usually called "MUX" for short. It selects binary information from one of many inputs and directs the output to a single line. The manner by which the output is selected is via a set of variables called selection inputs. The multiplexer or data selector is a derivative of a decoder. In general, a 2^n to 1 line MUX is derived from an n to 2^n decoder by adding 2^n input lines corresponding to each data input. The OR gate collates the outputs of the AND gates into a single output line. Mux may be implemented also using functional blocks which contain the basic gates inside.

Illustrative Problem # 12: Design a 4-to-1 MUX

Step 1 : There are four inputs.

Step 2 : Generate the Truth Table.

S1	S0	F
0	0	C0
0	1	C1
1	0	C2
1	1	C3

The select lines, S0 and S1 determines which line input will be selected for each input combination.

Skip Step 3

Step 4 : Draw the Logic Diagram and the equivalent functional block.

Logic Diagram :

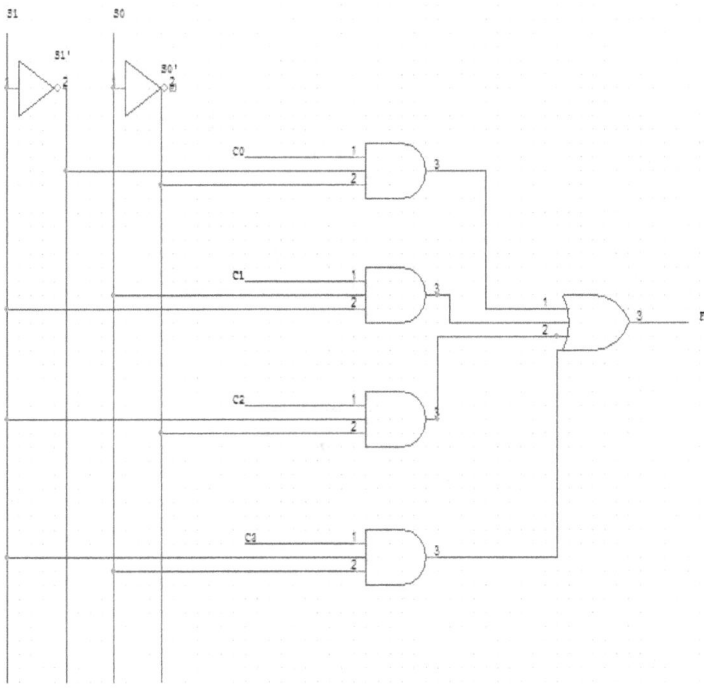

Functional Block:

This functional block contains the circuit above.

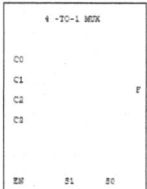

Illustrative Problem # 13: Design an 8-to-1 MUX

Step 1 : There are eighth inputs.

Step 2 : Generate the truth table

S2	S1	So	OUTPUT, F1
0	0	0	C0
0	0	1	C1
0	1	0	C2
0	1	1	C3
1	0	0	C4
1	0	1	C5
1	1	0	C6
1	1	1	C7

Skip Steps 3 and 4

Step 5 : Draw the 3-to-8 MUX Functional Block

```
        8 - TO - 1  MUX
   C7
   C6
   C5
   C4
                    F
   C3
   C2
   C1
   C0

   EN   S2   S1   S0
```

Illustrative Problem # 14: Implement the following Boolean Function with a multiplexer.

$$F(X,Y,Z) = \sum\nolimits_m(1,2,5,7)$$

Step 1 : There are three inputs, n=3.

The first n-1 variables are used as the selection variables, X and Y. The last variable, Z is used for the data input.

Step 2 : Generate the Truth Table and express F in terms of Z.

Data Input Line	X	Y	Z	F	
0	0	0	0	0	F = Z
	0	0	1	1	
1	0	1	0	1	F = Z'
	0	1	1	0	
2	1	0	0	0	F = Z
	1	0	1	1	
3	1	1	0	0	F = Z
	1	1	1	1	

Note that when F has the same value as Z in pairs, F = Z. For example when Z and F are equal to 0 and 1 respectively as in Data Input Line 0, F = Z. For the Data Input Line 1 when Z is 0 and 1 respectively, F = Z'. For instances wherein F assumes two consecutive zeroes, F =0. When it assumes two consecutive ones, F = 1.

Skip Step 3 and 4.

Step 5 : Draw the logic diagram using the functional block for a 4-TO-1 MUX.

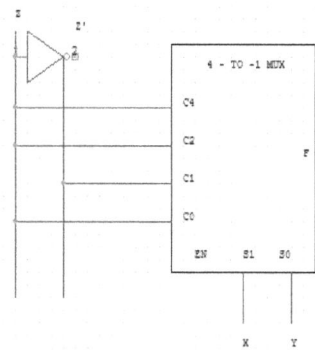

Illustrative Problem # 15 : Implement a Full Adder using a Multiplexer

Step 1: There are three inputs. Identify the Select and Input Lines.

   Number of Select Lines = n -1 = 3-1 =2 . Let X and Y be the Select Lines.

   There is one data input line. Let Z be the input line

   Input Combinations :

   00 = 0   01 = Z

   11 = 1   10 = Z'

Step 2 : Generate the Truth Table and express the output S and C in terms of Z.

Data Input Line	Index	X	Y	Z	S	C
0	0	0	0	0	0, S=Z	0, C=0
	1	0	0	1	1, S=Z	0, C=0
1	2	0	1	0	1, S=Z'	0, C=Z
	3	0	1	1	0, S=Z'	1, C=Z
2	4	1	0	0	1, S=Z'	0, C=Z
	5	1	0	1	0, S=Z'	1, C=Z
3	6	1	1	0	0, S=Z	1, C=1
	7	1	1	1	1, S=Z	1, C=1

Step 3 and 4. From Illustrative Problem #3

$$S(X,Y,Z) = \sum\nolimits_m(1,2,4,7)$$

$$C(X,Y,Z) = \sum\nolimits_m(3,5,6,7)$$

Step 5 : Since there are four possible inputs, the function is implemented using 4 x 1 MUX.

Illustrative Problem # 14 : Implement the following function using a four-input MUX

$$F(W,X,Y,Z) = \sum m\ (2,4,6,8,10,12,14)$$

Step 1 : There are 4 inputs, n=4. The number of selection lines is equal to n-1 or 4-1=3. The first three variables are used as the selection variables, W=S0, X=S1 and Y=S2. The last variable, Z is used for Data Input.

$$00 = 0 \qquad 11 = 1$$

$$01 = Z \qquad 10 = Z'$$

An OBE Approach to Logic Circuits and Digital Design: Step by Step

Step 2 : Generate the Truth Table

Data Input Line	Index	W	X	Y	Z	F	F
0	0	0	0	0	0	0	0
	1	0	0	0	1	0	
1	2	0	0	1	0	1	Z'
	3	0	0	1	1	0	
2	4	0	1	0	0	1	Z'
	5	0	1	0	1	0	
3	6	0	1	1	0	1	Z'
	7	0	1	1	1	0	
4	8	0	0	0	0	1	Z'
	9	1	0	0	1	0	
5	10	1	0	1	0	1	Z'
	11	1	0	1	1	0	
6	12	1	1	0	0	1	Z'
	13	1	1	0	1	0	
7	14	1	1	1	0	1	Z'
	15	1	1	1	1	0	

Skip Step 3 and 4.

Step 5. Draw the Logic Diagram

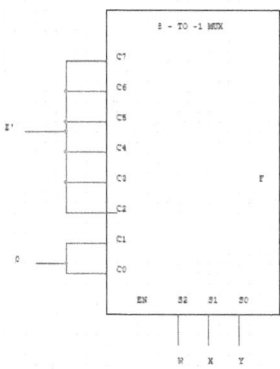

The three selection inputs, W,X,Y correspond to the Selection Inputs S2, S1 and S0 respectively. The truth table shows the value for the inputs. For example, when F1(W,X,Y) = 010, F1 = Z', so the complemented Z is applied to Data Input 2 as shown in the truth table.

The size of the MUX is determined by the number of the data input lines. Size of the MUX = $2^{n-1} = 2^{4-1} = 8$. So an 8 x 1 MUX is required.

DEMULTIPLEXER OR DEMUX

A DEMULTIPLEXER is the inverse of a multiplexer. It is a one-to-many circuit. The idea is for one input in conjunction with n select lines to select any of the m possible outputs. As opposed to a Multiplexer where the inputs are the ones transferred to the output. A careful examination of the demux circuit reveals that it is similar to a 2-to-4 decoder with enable.

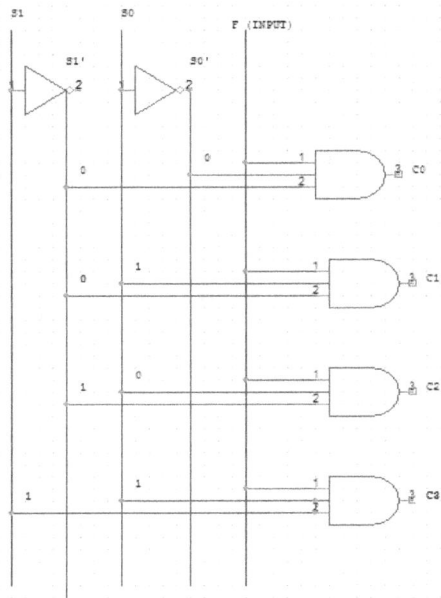

CHAPTER 8 END OF THE CHAPTER ACTIVITIES:

1. Compare the design of a BCD-to-Decimal Decoder with the unused combinations equal to zero and with the same unused combinations (10 to 15) as don't care combinations. Which circuit is better?

2. Implement the following Boolean functions with an 8-to-1 Multiplexer.

    2.1 $F(W,X,Y,Z) = \sum m\ (1,2,5,7,11,13,15)$

    2.2 $F(W,X,Y,Z) = \sum m\ (2,4,6,10,13,14)$

    2.3 $F(W,X,Y,Z) = \sum m\ (0,3,5,7,8,9,11,13)$

    2.4 $F(W,X,Y,Z) = \sum m\ (3,5,9,11,14,15)$

    2.5 $F(W,X,Y,Z) = \sum m\ (1,5,7,13,15)$

    2.6 $F(W,X,Y,Z) = \sum m\ (0,2,4,6,8,15)$

    2.7 $F(W,X,Y,Z) = \sum m\ (1,5,7,9,11,12,13)$

3. Repeat problem 2 using the appropriate Decoder.

4. How many 4-to-16 and 2-to-4 Decoders are required for a 6-to-64 Decoder.

5. Implement a Full Subtractor using a 3-to-8 decoder.

6. Use NAND gates to implement a 3-to-8 decoder.

7. Use Problem 6 to illustrate that it is equivalent to a two level AND-OR Circuit or an SOP function.

8. If the AND gates and INVERTERS in the MUX are like a decoder circuit, which component of the MUX differentiate it from a Decoder and makes it a MUX.

9. How many 3-to-8 and 2-to-4 decoders are required to construct a 5-to-32 Decoder. Draw the Logic Diagram.

10. How many 8-to-1 Line and 2-to-1 Line MUX are required to implement a 32-to-1 Line MUX?

STEP UP : FOR YOUR APPS

Decoder App

    Input : Higher Order Decoder

    Output : Number of Lower Order Decoder Required

        Logic Diagram

Multiplexer App

    Input : Boolean Function (SOP)

    Output : Logic Diagram

3. MUX App

    Input : Higher Order App

    Output : Number of Lower Order MUX Required

        Logic Diagram

EXPERIMENT 7 : CLC DESIGN – CODE CONVERTER

BCD TO SEVEN SEGMENT DECODER

Learning Outcomes :

1. To describe the packaging of the seven segment display

2. To compare the common anode with the common cathode seven segment display

3. To explain how the BCD-to Seven Segment Decoder drives the seven segment display

4. To verify the theoretical truth table generated during the design process.

Materials:

Digital Trainer or Breadboard, Power Supply, Logic Gates : 7445, 7446,7447,7448 or 7449 BCD to Seven Segment Decoder or Driver, Common Anode (CA) or Common Cathode (CC) Seven Segment Display, Connecting Wires, DATA SHEETS.

Procedure ;

Step 1 : Draw the pin configuration of the available Seven Segment Decoder . Refer to your Data Sheet.

Step 2 : Draw the pin configuration of the available Seven Segment Display. Refer to the Data Sheets. Is it Common Cathode or Common Anode ? What is the difference?

Diagram of the Seven Segment Display :

Step 3 : Combine Step 1 and Step 2 components. Draw the circuit. Using this experimental circuit, generate the Truth Table, illustrating the numbers 0 to 9 as the segments light up accordingly. Did all the output lights match the appropriate input. For example 0000 as 0, 0101 as 5, 1000 as 8, etc.?

_____
_____
_____
_____

Experimental Circuit :

Truth Table :

Step 4 : Compare the Generated Truth Table with the Theoretical Table. Are they the same ? Why ?

_____
_____
_____
_____

Step 5 : Movie Time – Record the video of your working circuit's simulation of numbers 0 to 9 and post online. List the link on the space below:

_____
_____

Step 6 : PICTURE, PICTURE! – Paste a picture of your team with your working circuit here.

CONCLUSION :

_____
_____
_____
_____

EXPERIMENT 8 : CLC DESIGN – ADDERS

FULL ADDER

Learning Outcomes:

1. To construct a full adder circuit.

2. To observe the operation of a full adder circuit.

3. To construct the full adder's truth table based on observation.

4. To compare the actual circuit's truth table with the theoretical.

Materials :

Digital Trainer or Breadboard, Power Supply, Logic Gates (7486- XOR, 7432 -OR , 7408- AND) , LEDS, series resistors, Connecting Wires, DATA SHEETS.

Procedure:

Step 1. Wire the Experimental Full Adder Circuit as shown . Remember to use different colors for the LED monitoring the carry and the sum .

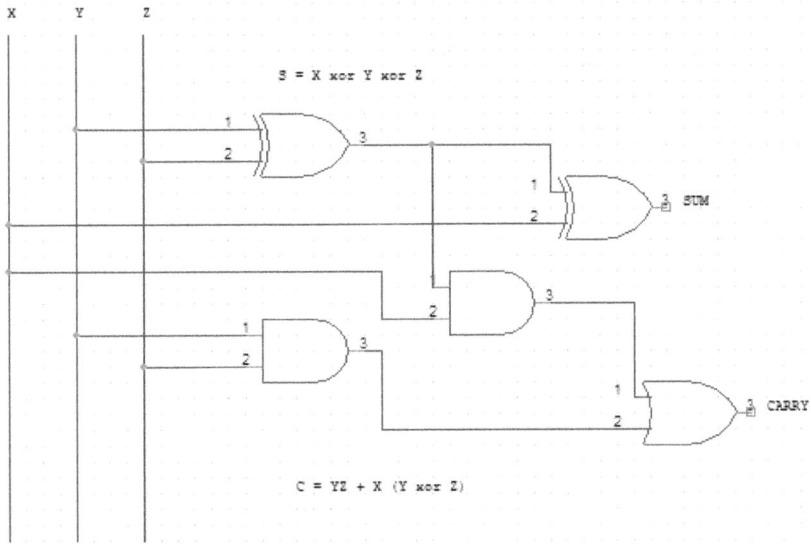

Step 2: PICTURE! PICTURE! – Paste a picture of your team and working circuit here:

Step 4: Generate the Truth Table based on actual inputs and LED status on the outputs. How many combinations?

Step 3 : Is the table on Step 2 the same as the Theoretical Truth Table of a Full Adder? Why?

_____
_____
_____
_____

CONCLUSION :

_____
_____
_____
_____
_____

EXPERIMENT 9 : CLC DESIGN – DECODERS : FULL ADDER IMPLEMENTED USING DECODERS

Learning Outcomes :

1. To show and prove that full adders can be implemented using decoders.

2. To describe the decoder chip packaging.

3. To construct the truth table of the actual circuit and compare with the theoretical.

4. To describe the different ways of implementing a full adder.

5. To recommend which is the best implementation among the many options.

Materials :

Digital Trainer or Breadboard, Power Supply, Integrated Circuits (2-to-4 Decoder, 3-to-8 Decoder, OR Gate), Power Supply, LEDS, series resistors, connecting wires. Data Sheets.

Procedure:

PART A – FULL ADDER USING 2-to-4 Decoders

Step 1 : Draw and Wire the experimental circuit while recalling to mind the following expressions:

$$S(X, Y, Z) = \sum m\ (1,2,4,7)$$
$$C(X, Y, Z) = \sum m\ (3,5,6,7)$$

Step 2 : Generate the truth table based on the actual circuit.

PART B – Full Adder using 3-to-8 Decoders.

Step 3. Wire the Experimental Circuit using an available 3-to-8 Decoder. Draw the Experimental Circuit Below.

Step 4. Generate the truth table based on the actual circuit.

Step 5. Are the Truth Tables in Steps 2 and 4 the same? Why?

_____
_____
_____
_____
_____
_____
_____

CONCLUSION:

_____
_____
_____
_____
_____
_____

EXPERIMENT 10 : CLC DESIGN –MULTIPLEXERS

FULL ADDER IMPLEMENTED USING MUX

Learning Outcomes:

1. To construct a full adder using multiplexer chip.

2. To describe the pin configuration of a MUX chip.

3. To explain how a MUX chip is wired.

4. To compare a full adder constructed with a MUX chip to a theoretical full adder.

5. To generate the truth table of an actual MUX full adder.

Materials :

Digital Trainer or Breadboard, Power Supply, MUX IC, LEDs, series resistor, OR gate, connecting wires, Data sheets.

Procedure :

Step 1 : Draw and wire a full adder made from a 4 to 1 MUX.

Step 2. Why is a 4 to 1 MUX used instead of an 8 to 1 ? Generate the truth table below based on the actual status of the output LEDs.

Step 3 : Compare the generated truth table with the theoretical truth table. Are they the same? Why?

Step 4 : PICTURE, PICTURE!: Paste a picture of your team with your Full Adder Circuit constructed from a Multiplexer below:

Conclusion:

_____
_____
_____
_____

CHAPTER 9 : BUILDING BLOCKS OF SEQUENTIAL LOGIC CIRCUIT

LEARNING OUTCOMES:

1. To describe the basic types of flip flops.

2. To explain how flip flop works.

3. To describe the different flip flop symbols.

4. To describe the Characteristic Table of the different types of flip flops.

SEQUENTIAL LOGIC CIRCUIT (SLC) is basically a Combinational Logic Circuit (CLC) with a storage or memory component. CLCs depend only at the input signal present when the output is taken. The output appears immediately when the input is applied. The output of the SLCs on the other hand is dependent on both present inputs and past outputs. The output signal is fed back to the input. It is a time-dependent circuit and stores the current state before it is excited again to produce the next state.

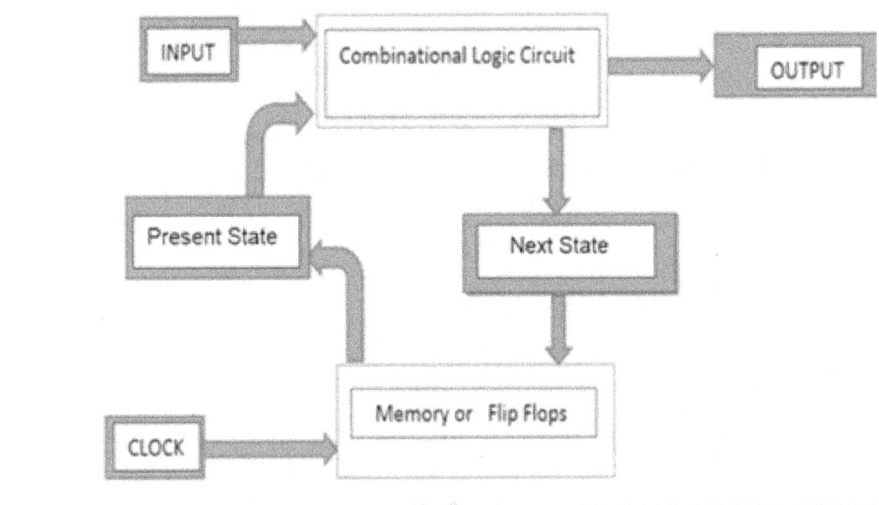

Figure 9.1 Block Diagram of a Sequential Logic Circuit

From Figure 9.1, it can be seen that a combinational logic circuit accepts the external inputs and the inputs (present state) coming from the storage and produces the signals for the external outputs and the next state which serves as the trigger for the memory circuit.

9.1 FLIP-FLOPS

The basic building block of a sequential circuit is a flip flop or a bistable multivibrator. It can store a 1-bit of memory until triggered to change states. As a binary memory circuit, it can store two values namely "1" or "0".

9.1.1 THE S-R (Set-Reset) Flip Flop

The SR flip flop has two inputs, Set and Reset and two outputs Q and Q' which are complements of each other. The operation of the SR flip flop is best described by simulating the output of its circuit equivalent as shown on Figure 9.3

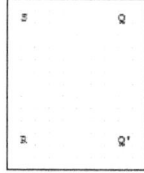

Figure 9.2. The SR Flip Flop Symbol

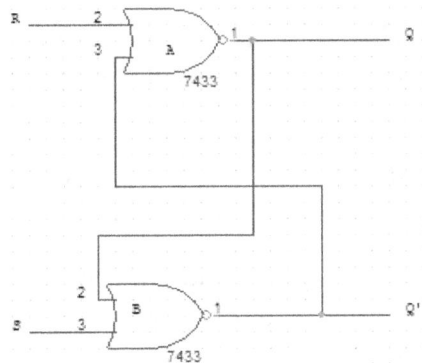

FIGURE 9.3 SET-RESET FLIP FLOP IMPLEMENTED USING NOR GATES

X	Y	F
0	0	1
0	1	0
1	0	0
1	1	0

Table 9.1 TRUTH TABLE FOR A NOR GATE

Table 9.2 Step by Step Simulation of the Operation of the SR Flip Fop

STEPS	S	R	Q(t+1)	Q(t+1)'	Remarks
1	0	0	Q(t)	Q(t)'	No Change
2	0	1	0	1	Reset
3	1	0	1	0	Set
4	1	1	0	0	Undefined
5	0	0			Indeterminate

We will simulate the operation of the Set Reset Flip Flop using Figure 9.3 and record our results, in a step by step fashion in Table 9.2.

Step 1 : S=0, R=0, No change. Therefore, the output Q(t+1) = Q(t). Q(t+1) is referred to as the Next State (t+1) and Q(t), the current or present state (t). If we apply 0 to S

Step 2. S=0, R=1. The 1 at the input of NOR gate A forces it to have an output of 0., Q(t+1) =0. Remember that the output of a NOR gate is 0 if any of the inputs is 1. If Q(t+1) = 0 and S=0, the output of Q(t+1)' = 1.

Step 3: S=1, R=0. At NOR gate A, Q(t+1) is 0 because Q(t+1)'=1 and R=0. At NOR gate B, we have at the input, Q(t+1) =1 and S=1, the output Q(t+1) therefore is 0.

Step 4: S=1, R=1, Q(t+1) =0, Q(t+1)'=1. At NOR gate A, we have 1 and 1 which gives us a Q(t+1)=1. At NOR gate B, we have, 0 and 1 at the input which gives us a Q(t+1)' =1. This is a forbidden or undefined state because Q(t+1) and Q(t+1)' must be complements of each other. Here they are both 1.

Step 5: If the previous state arises, both S and R will return to 0 at the same time and then each one will race to assume a state. Since this action cannot be predicted, this is labelled an indeterminate case.

CLOCKED SR FLIP FLOP

Synchronous circuits are time dependent. They change their states only with the appropriate timing signal. In general, it is dependent on the signal being high or low. A more precise triggering depends on the transition of the clock signal from high to low (negative edge-triggered) or from low to high (positive edge triggered). Figure 9.4, shows a clocked SR-flip flop based on NOR gates.

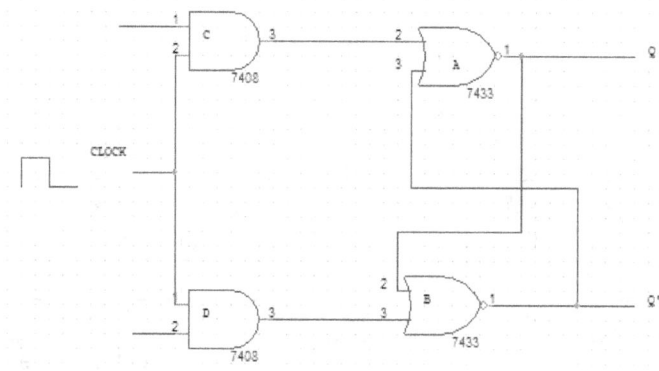

FIGURE 9.4. CLOCKED S-R FLIP FLOP

Note that for Figure 9.4 the output of the AND gate will be zero if the clock is 0. The input will be passed on to the output of the AND gate if and only if the clock is 1. Thus, both S and R are certain to be at the input of the respective NOR gates at the same time or in a synchronized fashion.

The characteristic table of a flip flop describes how a flip flop operates much like a truth table for logic gates.

TABLE 9.3 . Characteristic Table of a Clocked Set-Reset (SR) Flip Flop

S(t)	R(t)	Q(t+1)	Q(t+1)'	Remarks
0	0	Q(t)	Q(t)'	No change
0	1	0	1	Reset
1	0	1	0	Set
1	1	-	-	Undefined

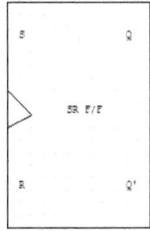

Figure 9.5 Clocked Set Reset Flip Flop Symbol

9.1.2 THE JK FLIP FLOP

This flip flop behaves like the Set Reset flip flop except that the output is defined when S=1 and R=1. The J input behaves like the Set (S) input and the K like the Reset(R) . When J=1 and K=1, the output is in the toggle state meaning the output switches to the complement state every time the clock passes.

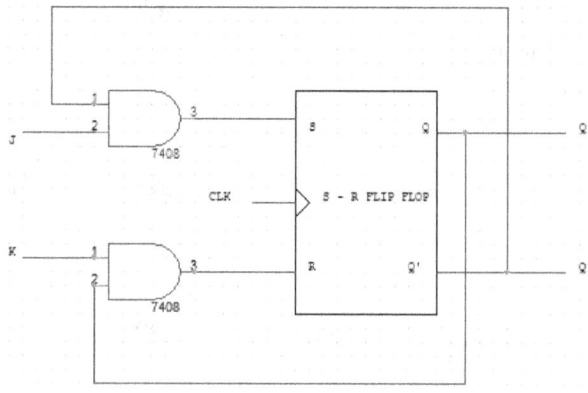

Figure 9.6 J-K Flip-Flop Implemented using SR Flip Flop

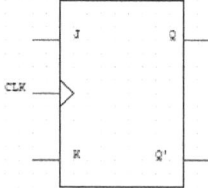

Figure 9.7 J-K Flip Flop Symbol

Table 9.4 Characteristic Table of a JK Flip Flop

J (t)	K(t)	Q(t+1)	Q(t+1)'
0	0	Q(t)	Q(t)'
0	1	0	1
1	0	1	0
1	1	Q(t)'	Q(t)

Table 9.5 Characteristic table to generate the Characteristic Equation of JK flip flop

J	K	Q(t)	Q(t+1)
0	0	0	0
0	0	1	1
0	1	0	0
0	1	1	0
1	0	0	1
1	0	1	1
1	1	0	1
1	1	1	0

$Q(t+1) = \sum m(1, 4, 5, 6)$

To come up with the equation we simplify the SOP output using K-map. The simplified function is :

$Q(t+1) = JQ(t)' + K'Q(t)$

9.1.3 THE T-FLIP FLOP

The T or Toggle flip flop can be constructed from a JK flip flop by tying the J and the K inputs.

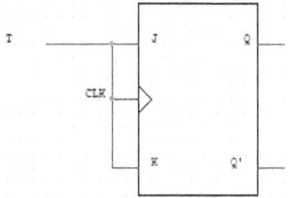

Figure 9.9 T flip flop symbol

Table 9.6 Characteristic table of a T Flip Flop

T	Q(t)	Q(t+1)
0	0	0
0	1	1
1	0	1
1	1	0

The characteristic equation after k-map is :

$Q(t+1) = TQ(t)' + T'Q(t)$

9.1.4 The D Flip Flop

The D (Delay) flip flop has only one input and the usual two outputs, Q and Q' which are complements of each other. The characteristic table is as shown at Table 9.10 and the logic symbol on Figure 9.10.

Table 9.10 Characteristic Table of a D Flip Flop

D(t)	Q(t+1)
0	0
1	1

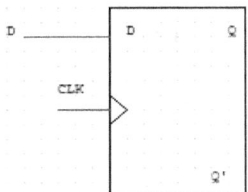

Figure 9.10 . D flip flop symbol

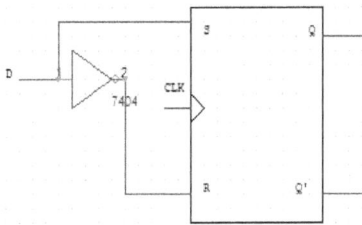

Figure 9.11 D flip flop implemented using SR flip flop

9.1.5 FLIP FLOP TRIGGERING

Synchronous or clocked flip flops maybe level-triggered or edge-triggered. Level triggering means the flip flop is triggered when the clock is either HIGH or LOW. A positive level triggered flip flop is triggered when the clock is HIGH. On the other hand, a negative level triggered flip flop is triggered when the clock is LOW.

Edge triggering means the triggering occurs during the transition from LOW to HIGH or HIGH to LOW. Positive Edge or Leading Edge Triggering occurs during the transition of the clock pulse from LOW to HIGH. On the other hand, Negative Edge or Trailing Edge Triggering occurs during the transition of the clock pulse from HIGH to LOW.

Figure 9.12 Positive Edge , Rising Edge or Leading Edge Triggering

CHAPTER 9 END OF THE CHAPTER ACTIVITIES

1. Implement an SR flip flop using NAND gates.

2. Implement a JK Flip Flop using NAND gates.

3. Illustrate positive edge and negative edge on a positive pulse.

4. Illustrate the leading and trailing edge on a negative pulse.

5. What is a Master Slave flip flop. Illustrate and generate the characteristic table.

6. Implement a T flip flop using NAND gates.

7. Prove using K-maps that $Q(t+1) = D(t)$ for the D flip flop.

Experiment # 11 : BASIC FLIP FLOP OPERATION

Learning Outcomes:

1. To observe the operation of the D flip flop on a simulator.
2. To observe the operation of the T flip flop on a simulator.
3. To observe the operation of a JK flip flop on a simulator.
4. To observe the operation of an SR flip flop on a simulator.
5. To compare these observations with the flip flops characteristic tables.

Materials :

Simulation Software, Data Sheets, Smartphone with Camera, Printer

Procedure :

Step 1 : Given an available Simulation Software. Placed the flip flop on the simulation canvass. Apply the clock and the input signals.

Step 2 : Trace the output and display with the input and clock . Take the screen shot and paste below.

Step 2.a D flip flop

Step 2.b T flip flop

Step 2.c JK flip flop

Step 2.d SR flip flop

Step 3 : Analyze the trace for Step2 . Are they consistent with the characteristic table of the flip flops? Why ?

___For D flip flop :_____

_____
_____
_____

___For T flip flop: _____

_____
_____

___For JK flip flop: _____

_____
_____
_____

___ For SR flip flop: _____

_____
_____
_____

CONCLUSION :

_____
_____
_____
_____
_____

CHAPTER 10 : THE ANALYSIS OF OF SEQUENTIAL LOGIC CIRCUITS

Learning Outcomes :

1. To describe the step by step process of analyzing sequential logic circuits.

2. To explain the tools of the analysis process.

3. To describe how the Excitation Tables of Flip Flops are used in State Tables.

The ANALYSIS of sequential logic circuit starts with a logic diagram and ends up with a state table and a state equation. Let us recall the block diagram of a Sequential Logic Circuit below (Figure 10.1) which we will use in the analysis process.

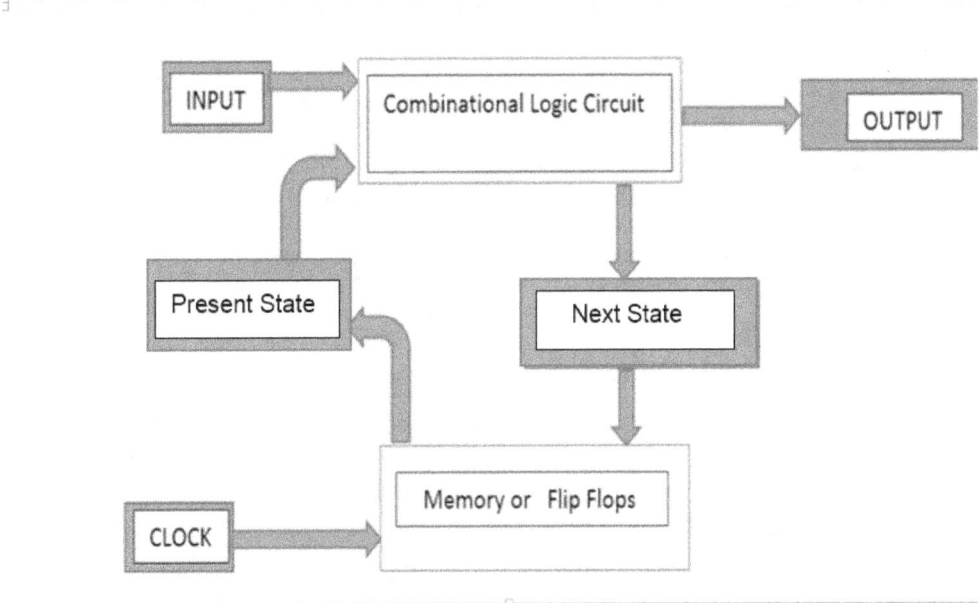

Figure 10.1 Block Diagram of a Sequential Logic Circuit

From the diagram, we can see that there are five signals involved, namely, clock, external input, external output, present state and next state signals. There are two main sources and destination of signals which are the flip flops and the combinational logic circuit. The external inputs and outputs are the sole signals for a combinational CLC. However in an SLC, the CLC receives another input which is the next state and another output which is the present state of the flip flop. We will use this block diagram to identify the components of every sequential logic circuit that we will analyze.

ANALYSIS PROCEDURE: STEP BY STEP

Step 1: Generate input equations of each flip flop in the circuit

Step 2: Generate Next State and Output Equations. Consider the Characteristic Table of the flip flop in generating the Next State Equations. We use this step only for D flip flop which has the same next state equation as the present state. For all other flip flops, proceed to step 3.

Step 3. Generate the State Table. The table lists the operational relationships between the five signals in the circuit. To generate the table, first list all the possible combinations of the present state and the inputs. The next state values are then derived from Step 2.

Step 4 : Generate the State Diagram.

Illustrative Problem # 1: Analysis of CLCs using D flip flops

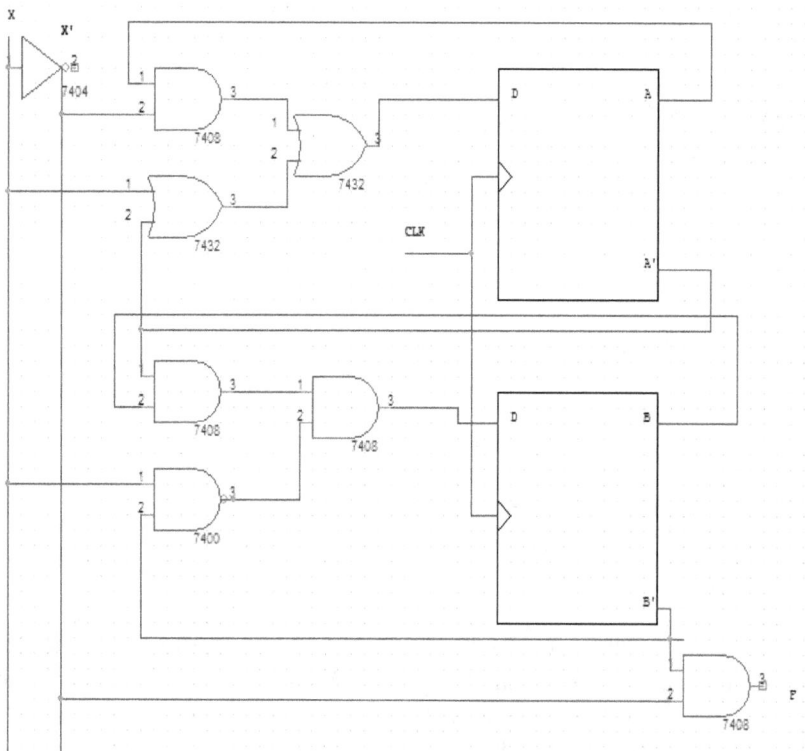

Step 1 : Generate the Flip Flop Input Equations

$DA = X'A + (X+A')$

$DB = (A'B)(XB')$

$F = B'X'$

Step 2 : Generate the Next State and Output Equations. The Next State Equations can be generated directly from the Flip Flop Input Equations. For this example using D flip flops, $A(t+1) = DA$ and $B(t+1)=B$. Recall that from the Characteristic Table of the D flip flop, $Q(t+1) = D$. Therefore the equations for the two flip flops and the output are :

$A(t+1) = DA = X'A + (X+A')$

$B(t+1) = DB = (A'B)(XB')$

$F = B'X'$

Step 3 : Generate the State Table.

A	B	X	A(t+1)	B(t+1)	F
0	0	0	1	0	1
0	0	1	1	0	0
0	1	0	1	0	0
0	1	1	1	0	0
1	0	0	1	0	1
1	0	1	1	0	0
1	1	0	1	0	0
1	1	1	1	0	0

Step 4 : Generate the State Diagram

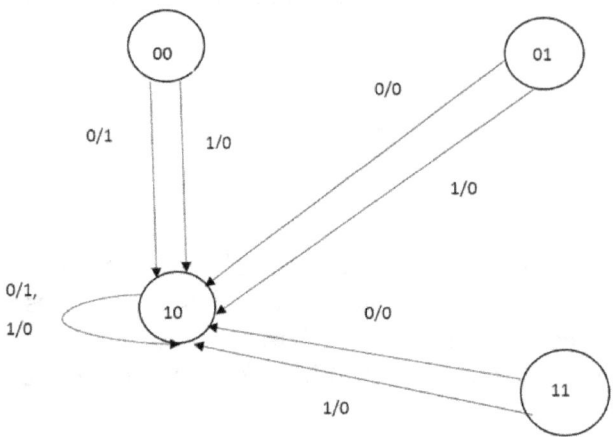

The analysis is complete with the generation of the Next State Table depicting the relationship of the input, output and state equations for all the possible combinations and the drawing of the state diagram.

Illustrative Problem # 2: ANALYSIS USING JK FLIP FLOPS

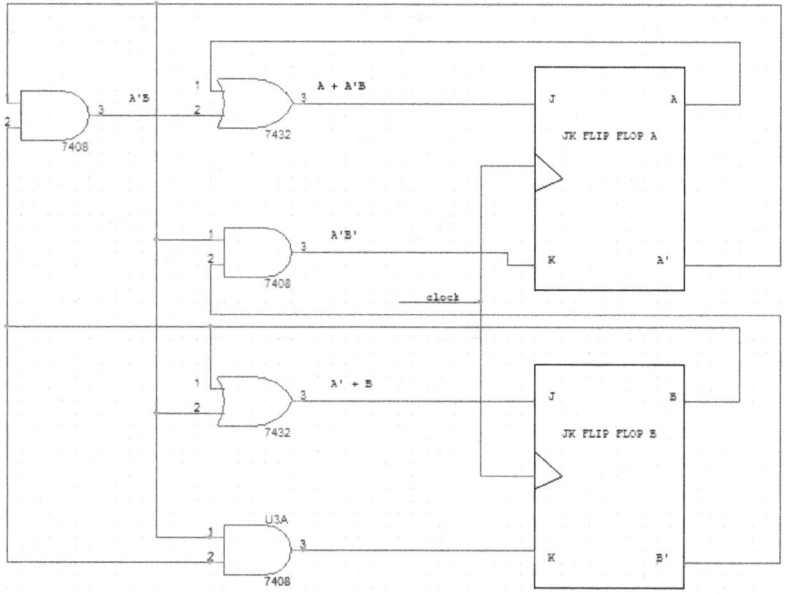

Step 1 : Generate the Flip Flop Input Equations

JA = A + A'B

KA = A'B'

JB = A' + B'

KB = A'B

Step 3 : Generate the State Table.

Characteristic Table of JK Flip Flop

J	K	Q(t+1)	Remarks
0	0	Q(t)	No change
0	1	0	Reset
1	0	1	Set
1	1	Q(t)'	Complement

Step 3.1 Find the values of each of the flip flop input equation of Step 1 in terms of the present state and input variables.

Step 3.2 The next state is then derived by referring to the characteristic table of the JK Flip Flop

A	B	JA	KA	JB	KB	A(t+1)	B(t+1)
0	0	0	1	1	0	0	1
0	1	1	0	1	1	1	0
1	0	1	0	1	0	1	1
1	1	1	0	0	0	1	1

An OBE Approach to Logic Circuits and Digital Design: Step by Step

Step 4 : Generate the state diagram

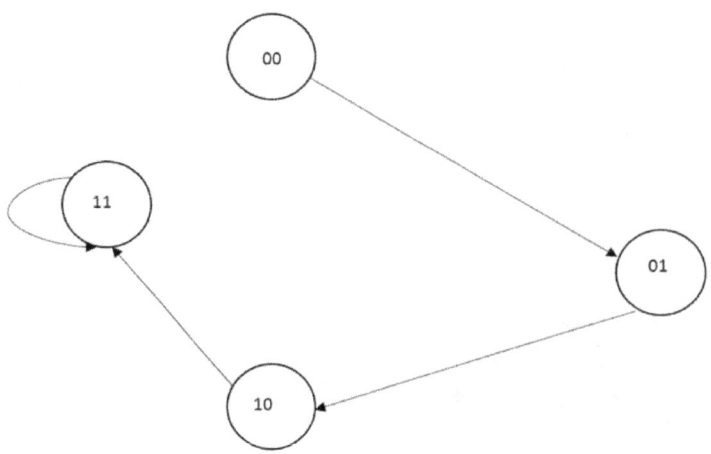

Illustrative Problem # 3 : SLC ANALYSIS OF T FLIP-FLOP

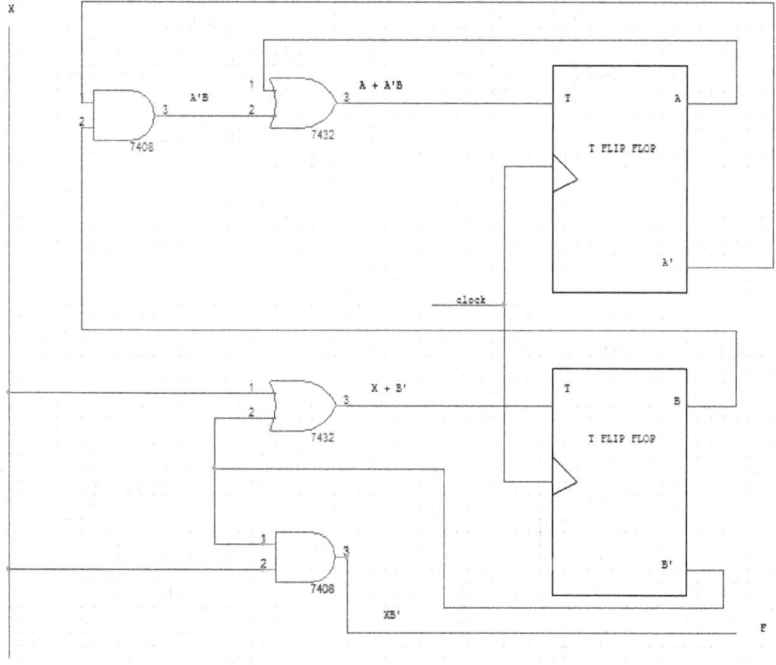

Step 1 : Generate the flip flop input equations

TA = A + A'B

TB = X + B'

F = XB'

Step 3 : Generate the State Table.

Characteristic Table of a T Flip-Flop

T	Q(t+1)	Remarks
0	Q(t)	No Change
1	Q(t)'	Toggle

Step 3.1 Find the values of each of the flip flop input equation of Step 1 in terms of the present state and input variables.

Step 3.2 The next state is then derived by referring to the characteristic table of the T Flip Flop

A	B	X	TA	TB	A(t+1)	B(t+1)	F
0	0	0	0	1	0	1	0
0	0	1	0	1	0	1	1
0	1	0	1	0	1	1	0
0	1	1	1	1	1	0	0
1	0	0	1	1	0	1	0
1	0	1	1	1	0	1	1
1	1	0	1	0	0	1	0
1	1	1	1	1	0	0	0

Step 4 : Draw the State Diagram

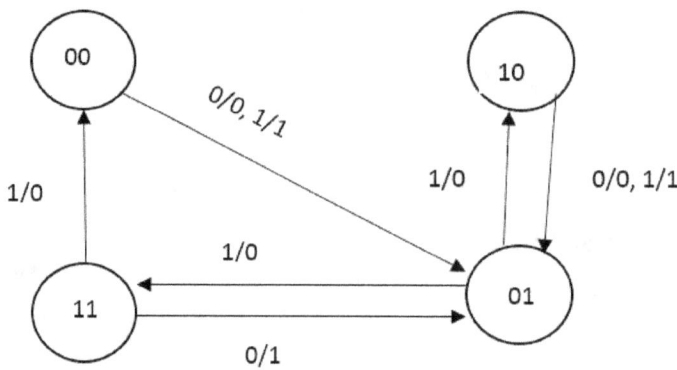

IP # 4 : Analyze the D flip flop below. Consider only the first 10 states.

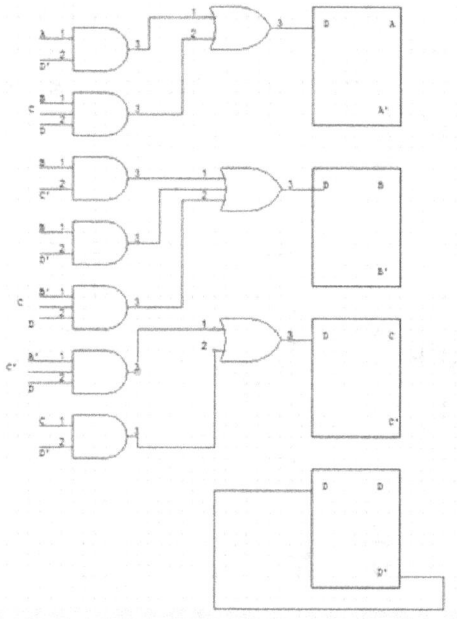

Step 1 : Generate the D flip flop input equation

$D_A = AD' + BCD$

$D_B = BC' + BD' + B'CD$

$D_C = A'C'D + CD'$

$D_D = D'$

Step 2 : Generate the Next State Equation. For the D flip flop, the flip flop input equations and the next state equations are the same.

$A(t+1) = D_A = AD' + BCD$

$B(t+1) = D_B = BC' + BD' + B'CD$

$C(t+1) = D_C = A'C'D + CD'$

$D(t+1) = D_D = D'$

Step 3 : Generate the State Table. A close analysis of the transition from the present state to the next state reveals that this is a counter that counts up from 0 to 9 and then resets.

Count	A	B	C	D	A(t+1)	B(t+1)	C(t+1)	D(t+1)
0	0	0	0	0	0	0	0	1
1	0	0	0	1	0	0	1	0
2	0	0	1	0	0	0	1	1
3	0	0	1	1	0	1	0	0
4	0	1	0	0	0	1	0	1
5	0	1	0	1	0	1	1	0
6	0	1	1	0	0	1	1	1
7	0	1	1	1	1	0	0	0
8	1	0	0	0	1	0	0	1
9	1	0	0	1	0	0	0	0

Step 4 : Generate the State Diagram.

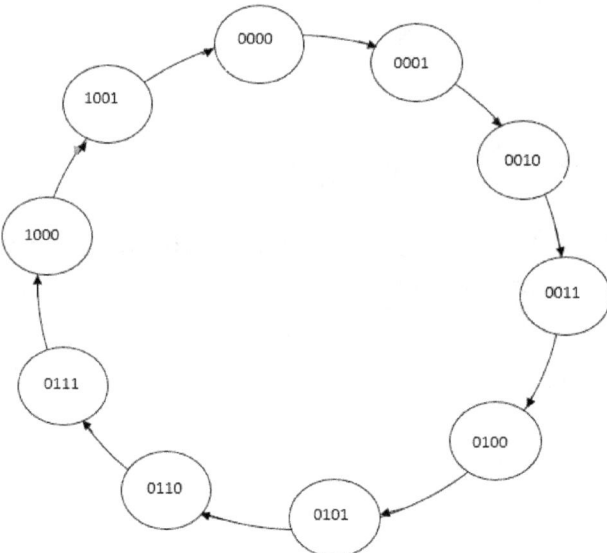

CHAPTER 10 END OF THE CHAPTER ACTIVITIES:

ANALYZE THE FOLLOWING CIRCUITS :

1.TWO D FLIP FLOP SLC, NO EXTERNAL INPUT NO OUTPUT. HOW MANY POSSIBLE COMBINATIONS?

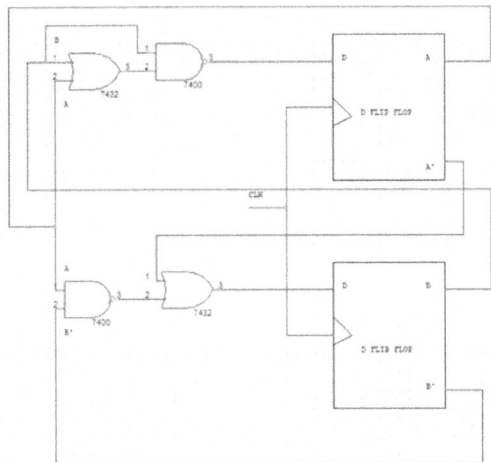

1.Add an external input, X . How many possible combinations? Remove the A inputs and connect to X, Analyze the circuit.

1.1 Will the addition of an output , F affect the number of possible combinations ?Why?

1.2.1 Repeat Problem 1 using T Flip Flops

1.2.2 What is the difference between problem 1 and problem 2. When will the output of problem 1 be the same with problem 2? Why?

1.2.3 Where is a possible application of this Circuit? What modification must you do to make it a circuit that you can identify.

1.2.4 Differentiate the application between the D and the T flip flops.

2. TWO JK FLIP FLOPS. ANALYZE THE CIRCUIT.

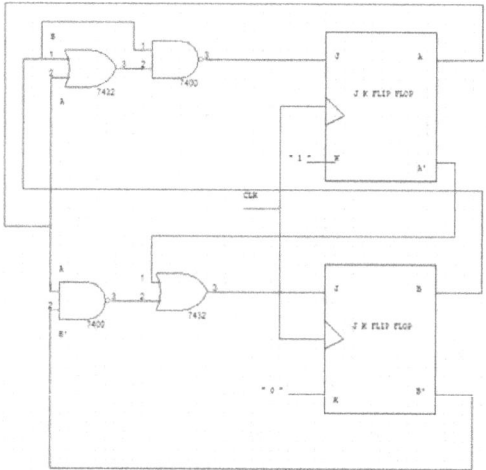

a. How many combinations? What determines the number of combinations, the number of flip flops or the number of inputs of the flip flops? Will the number of external inputs affect the number of possible combinations?

b. Does two D flip flop and two JK flip flops with no external inputs have the same number of possible combinations?

c. What is a possible application of a JK flip flop ?

3.TWO D Flip Flops, 1 input and 1 output SLC

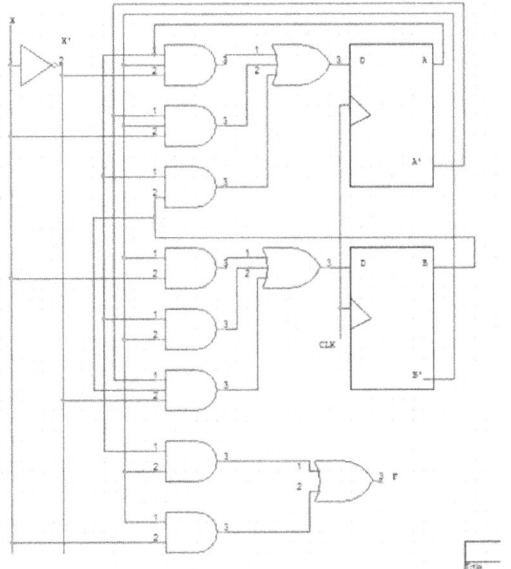

4. Two T flip flops, 1 input, 1 output SLC

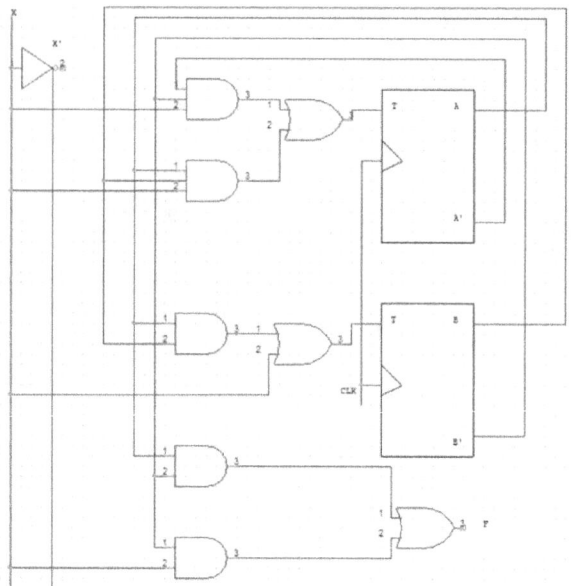

5.

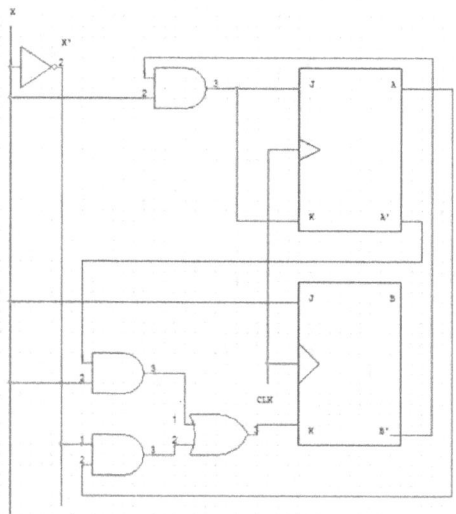

EXPERIMENT 12 : THE ANALYSIS OF A D FLIP FLOP CIRCUIT

Learning Outcomes:

1. To analyze a D flip flop circuit.

2. To produce the appropriate input, clock, present state, next state and output signals on the available simulator.

3. To interpret the resulting trace and compare with the theoretical.

Materials :

Simulation Software, Data Sheets, Smartphone with Camera, Printer

Procedure :

Step 1 : Draw on the available simulator canvass the given D flip flop circuit :

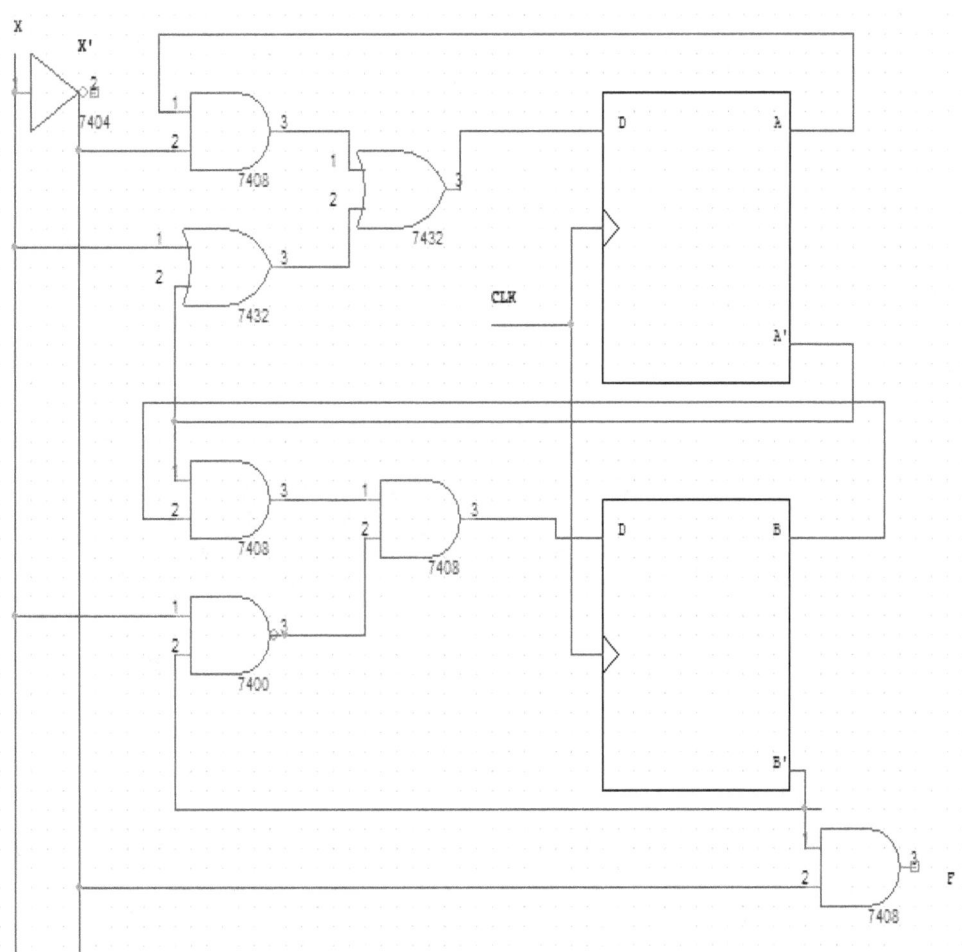

Step 2 : Apply the supply ,clock, the inputs and all the necessary signals. Monitor the output.

Step 3 : Create a trace for the input the clock and the output and paste the screenshot below.

For example :

CLK

X

D_A

D_B

A

B

F

Step 4 : Generate the state table of the circuit.

Is the screenshot consistent with the state table of the circuit. Why?

_____
_____
_____
_____
_____

Conclusion:

_____
_____
_____

EXPERIMENT 13 : THE ANALYSIS OF A T FLIP FLOP CIRCUIT

Learning Outcomes:

1. To analyze a T flip flop circuit using a simulation software.

2. To produce the appropriate input, clock, present state, next state and output signals on the available simulator.

3. To interpret the resulting trace and compare with the theoretical.

Materials:

Simulation Software, Data Sheets, Smartphone with Camera, Printer

Procedure:

Step 1 : Draw on the available simulator canvass the T flip flop circuit below :

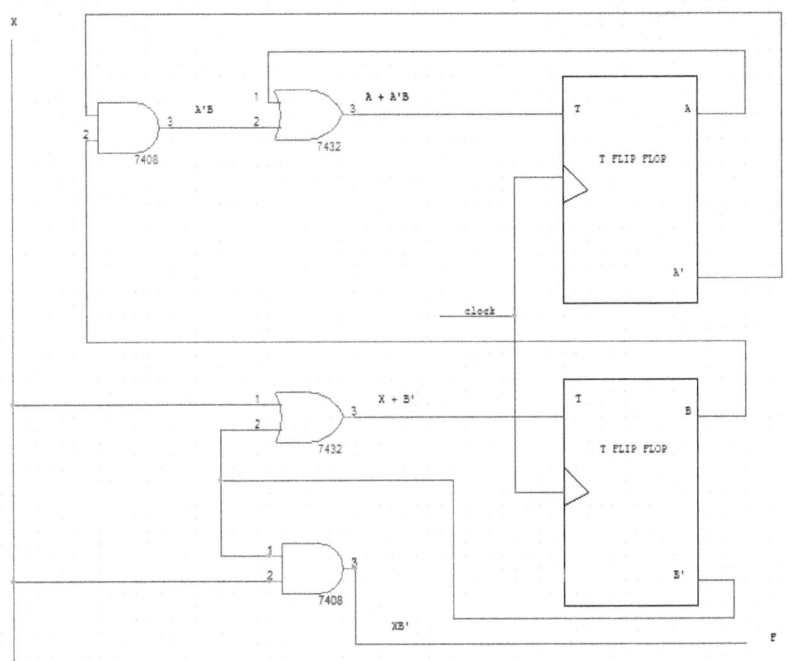

Step 2 : Apply the supply ,clock, the inputs and all the necessary signals. Monitor the output.

Step 3 : Create a trace for the input the clock and the output and paste the screenshot below.

For example :

CLK

X

T_A

T_B

A

B

F

Step 4 : Generate the state table of the circuit.

Step 5 : Compare the screenshot with the state table. Is it consistent with the state table of the circuit. Why?

_____
_____
_____
_____

Conclusion:

_____
_____
_____
_____
_____

EXPERIMENT 14 : THE ANALYSIS OF A JK FLIP FLOP CIRCUIT

Learning Outcomes :

1. To analyze a T flip flop circuit using a simulation software.

2. To produce the appropriate input, clock, present state, next state and output signals on the available simulator.

3. To interpret the resulting trace and compare with the theoretical.

Materials :

Simulation Software, Data Sheets, Smartphone with Camera, Printer

Procedure :

Step 1 : Draw on the available simulator canvass the JK flip flop circuit below :

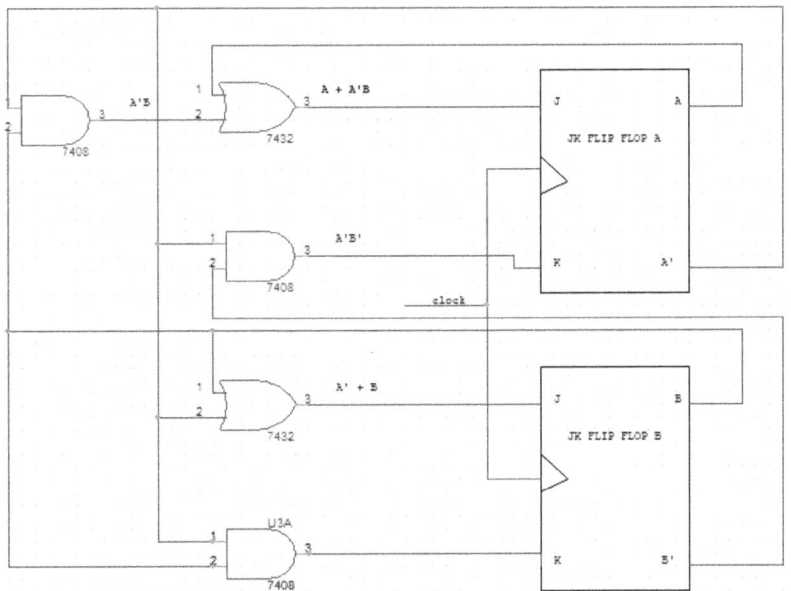

Step 2 : Apply the supply ,clock, the inputs and all the necessary signals. Monitor the output.

Step 3 : Create a trace for the input the clock and the output and paste the screenshot below.

For example :

CLK

J_A

K_A

J_B

K_B

A

B

Step 4 : Generate the state table of the circuit.

Step 5 : Compare the screenshot with the state table. Is it consistent with the state table of the circuit. Why?

_____
_____
_____
_____
_____

Conclusion:

_____
_____
_____
_____
_____

CHAPTER 11 : THE DESIGN OF SEQUENTIAL LOGIC CIRCUITS

LEARNING OUTCOMES :

1. Demonstrate the skill of designing a Sequential Logic Circuit (SLC)

2. Identify and use the tools for designing SLCs.

3. Explain how state tables, state equations and state diagrams are interrelated.

4. Explain the link between analysis and design.

The design process starts with the given specifications in the form of a worded description of the function of the SLC, a state equation or state diagram and ends with a logic diagram.

STEP BY STEP PROCEDURE :

1. From the given worded specification, generated the state diagram identifying the present state and next state values and the input/output values for each transition.

2. Generate the State Table detailing the binary values for the inputs, present state, next state and outputs. The number of combinations is determined by the number of inputs and the number of flip flops. The number of flip flops is equivalent to the number of bits inside the state diagram bubble.

3. Derive the Flip Flop Input Equations by following the steps below:

    3.a For D Flip Flops : Derive the Next State Equation in SOP abbreviated form

        3.a.1. Simplify using Karnaugh. The Next State Equation and the Flip Flop Input Equations are the same.

    3.b For T, SR and JK Flip Flops: From the Next State Values, get the Flip Flop Input Equations by referring to the Excitation Table of the Flip Flop.

4. Draw the logic diagram.

FLIP FLOP EXCITATION TABLES

For the design process, the flip flop excitation table is used. It lists the output firsts then the inputs that will result to these outputs. In the design process we begin with the output in mind and then we trace back to the logic diagram that will result to this output.

Table 10.1 Excitation Table of D Flip Flop

Q(t)	Q(t+1)	D
0	0	0
0	1	1
1	0	0
1	1	1

Table 10.2 Excitation table for T Flip-Flop

Q(t)	Q(t+1)	T
0	0	0
0	1	1
1	0	1
1	1	0

Table 10.3 Excitation Table for SR Flip Flop

Q(t)	Q(t+1)	S	R
0	0	0	X
0	1	1	0
1	0	0	1
1	1	X	0

Table 10.4 Excitation table for JK

Q(t)	Q(t+1)	J	K
0	0	0	X
0	1	1	X
1	0	X	1
1	1	X	0

An OBE Approach to Logic Circuits and Digital Design: Step by Step

Illustrative Problem # 1: Design an SLC with the State Diagram Below

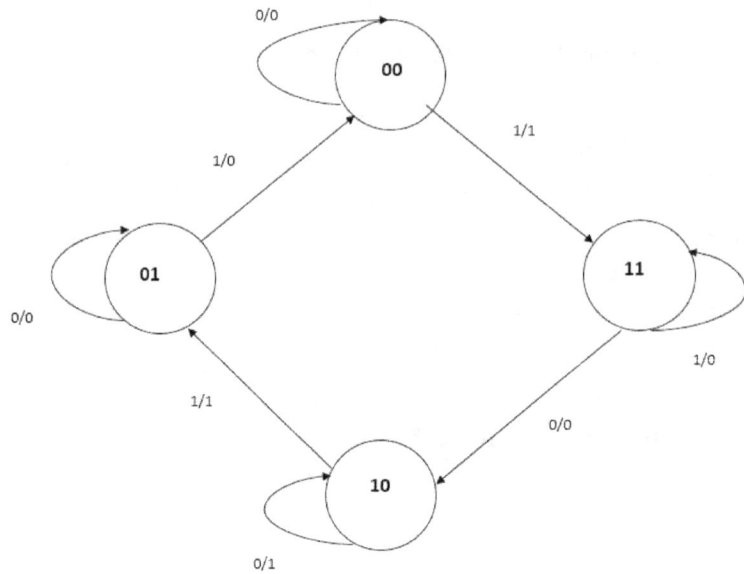

Step 2 : Generate the State Table. From the state diagram, there is one input and two flip flops which means there are $2^3 = 8$ possible combinations.

PRESENT STATE		INPUT	NEXT STATE		OUTPUT
A(t)	B(t)	X(t)	A(t+1)	B(t+1)	F
0	0	0	0	0	0
0	0	1	1	1	1
0	1	0	0	1	0
0	1	1	0	0	0
1	0	0	1	0	1
1	0	1	0	1	1
1	1	0	1	0	0
1	1	1	1	1	0

Step 3.a Derive the Next State Equations and the output from the table in SOP abbreviated form.

$D_A = A(t+1) = \sum_m (1,4,6,7)$

$D_B = B(t+1) = \sum_m (1,2,5,7)$

$F = \sum_m (1,4,5)$

Step 3. a.1 Simplify using Karnaugh Map

The simplified functions are :

DA = AB'X' + A'B'X + AB

DB = B'X + AB' + A'BX'

$\qquad$ F = AB' + B'X

Step 4 : Draw the Logic Diagram

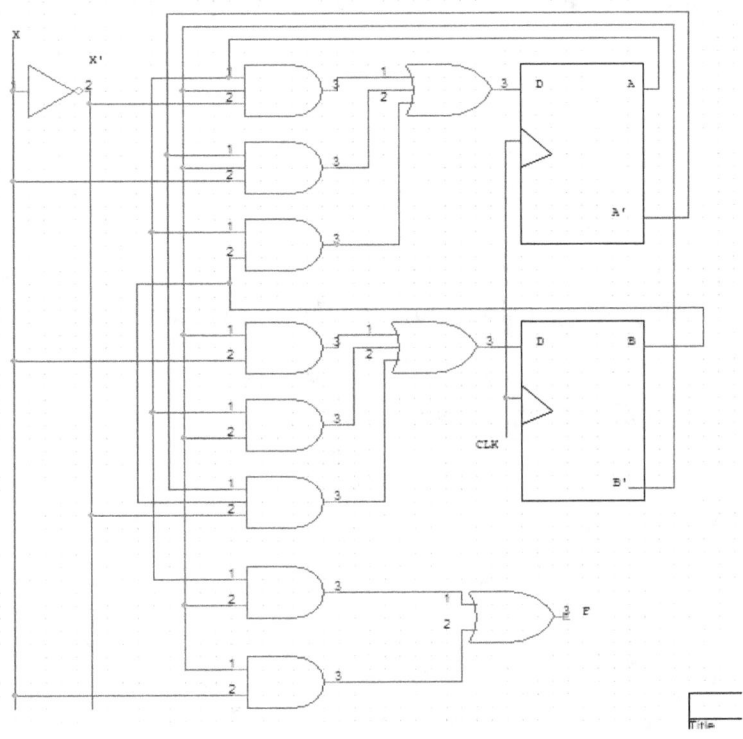

Illustrative Problem # 2 : Repeat IP#1 using T Flip Flops.

Step 2 : Generate the State Table. Note that an additional columns will be added for input equations of the T Flip Flops. The T flip flops input columns are filled up by referring to Table 10.2 T Flip flop Excitation table. Note that if there is a change from present state value to next state value, the value of T is 1 (toggle). If there is no change from present to next state T = 0 (no change)

Present State		Input	Next State		Output	T Flip Flop Inputs	
A(t)	B(t)	X(t)	A(t+1)	B(t+1)	F	T_A	T_B
0	0	0	0	0	0	0	0
0	0	1	1	1	1	1	1
0	1	0	0	1	0	0	0
0	1	1	0	0	0	0	1
1	0	0	1	0	1	0	0
1	0	1	0	1	1	1	1
1	1	0	1	0	0	0	1
1	1	1	1	1	0	0	1

Step 3.b Generate the T Flip Flop Input Equations from the table in SOP abbreviated form. The output, F is still the same as IP#1. It is not affected by changes in the F/F because it is F/F independent.

$T_A = \sum_m (1,6)$

$T_B = \sum_m (1,3,5,6,7)$

$F = \sum_m (1,4,5)$

Step 3.b.1 Simplify the Equations using Karnaugh Map. (Maps in the notebook)

The following are the simplified functions :

TA = A'B'X + ABX'

TB = X + AB

F = AB' + B'X

Step 4 . Draw the Logic Diagram

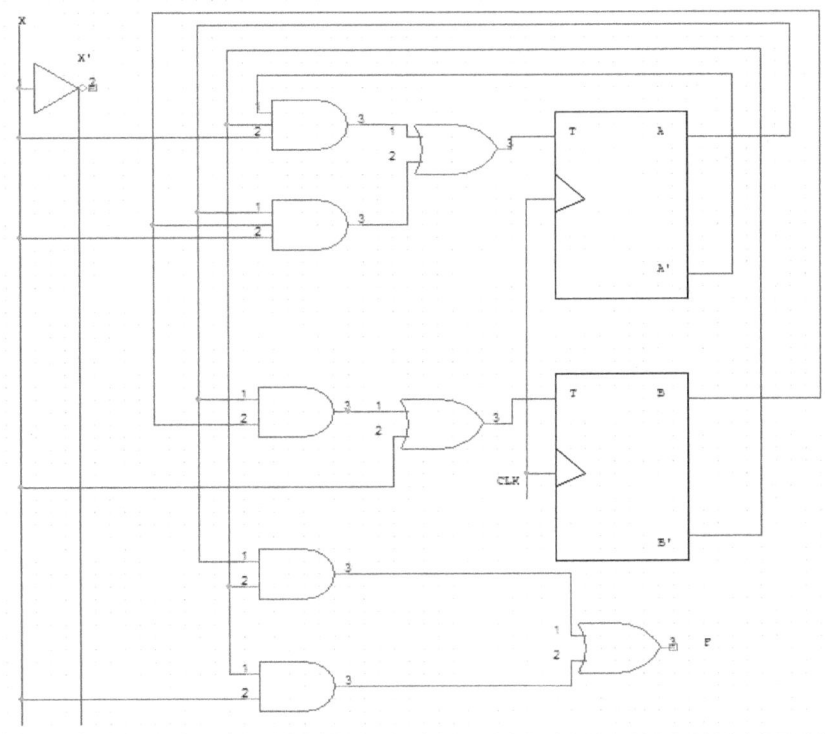

Illustrative Problem # 3 : Repeat IP # 1 using JK flip flops

Step 2 : Generate the State Table. The JK Flip Flop Input Equation Values are generated by referring to Table 10.4, Excitation table for JK flip flops.

Present State		Input	Next State		Output	JK Flip Flop Inputs			
A	B	X	A(t+1)	B(t+1)	F	JA	KA	JB	KB
0	0	0	0	0	0	0	X	0	X
0	0	1	1	1	1	1	X	1	X
0	1	0	0	1	0	0	X	X	0
0	1	1	0	0	0	0	X	X	1
1	0	0	1	0	1	X	0	0	X
1	0	1	0	1	1	X	1	1	X
1	1	0	1	0	0	X	0	X	1
1	1	1	1	1	0	X	0	X	0

Step 3.a Generate the JK Flip Flop Input Equations from the table above in SOP abbreviated form.

JA = $\sum$m (1), d(4,5,6,7)

KA = $\sum$m (5), d(0,1,2,3)

JB = $\sum$m(1,5), d(2,3,6,7)

KB = $\sum$m(3,6), d(0,1,4,5)

Step 3.b . Simplify using Karnaugh maps

The following are the simplified functions:

JA = B'X

KA = B'X

JB = X

KB = A'X + AX'

Step 4 : Draw the logic diagram

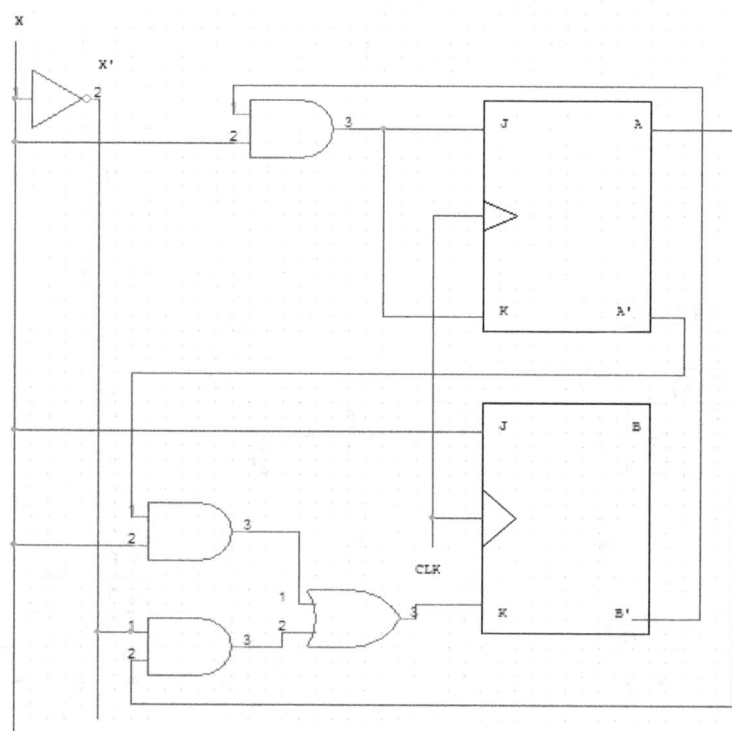

Illustrative Problem # 4 : Design a counter that counts up from 0 to 9 and then resets. Use D flip flops.

Solution :

Step 1 : Draw the state diagram from the worded specification. The worded problem specifies to count up therefore each bubble next to each other is incremented by 1. The number of bits inside the bubble is derived from the highest count equal to 9. Recall that $N = 2^n$, where n is equivalent to the number of bits, with $2^4 > 9$, n= 4. Three bits will not satisfy the requirement because $2^3 = 8$. There are four (n=4) flip flops.

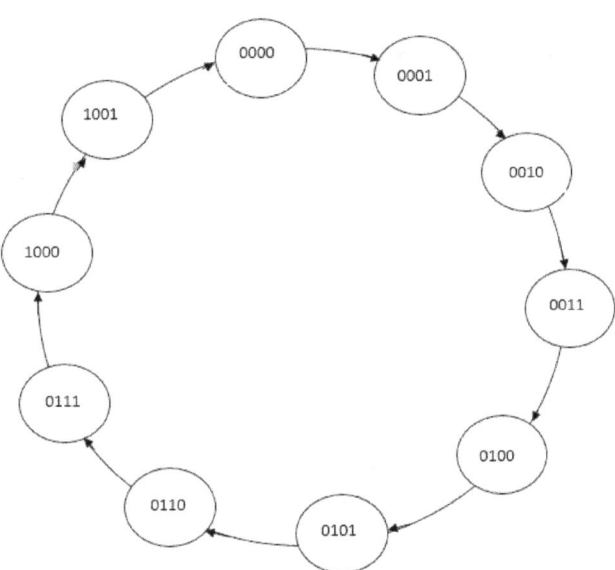

Step 2 : Generate the State Table

Count	A	B	C	D	A(t+1)	B(t+1)	C(t+1)	D(t+1)
0	0	0	0	0	0	0	0	1
1	0	0	0	1	0	0	1	0
2	0	0	1	0	0	0	1	1
3	0	0	1	1	0	1	0	0
4	0	1	0	0	0	1	0	1
5	0	1	0	1	0	1	1	0
6	0	1	1	0	0	1	1	1
7	0	1	1	1	1	0	0	0
8	1	0	0	0	1	0	0	1
9	1	0	0	1	0	0	0	0
10 to 15	Don't	Cares						

Step 3.a : Generate the Next State Equations :

$A(t+1) = D_A = \sum m\ (7,8), d\ (10\ to\ 15)$

$B(t+1) = D_B = \sum m\ (3,4,5,6), d(10\ to\ 15)$

$C(t+1) = D_C = \sum m\ (1,2,5,6), d(10\ to\ 15)$

$D(t+1) = D_D = \sum m\ (0,2,4,6,8), d(10\ to\ 15)$

Step 3.a.1 Simplify using K-maps .

The following are the simplified functions :

DA = AD' + BCD

DB = BC' + BD' + B'CD

DC = A'C'D + CD'

DD = D'

Step 4 : Draw the Logic Diagram

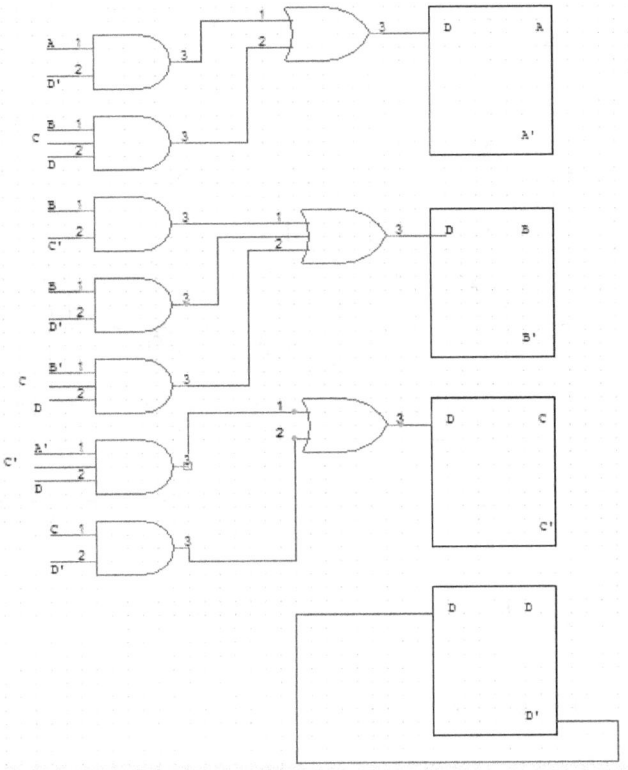

An OBE Approach to Logic Circuits and Digital Design: Step by Step

IP# 5 : Repeat IP#4 using T flip flops

Step 2 : Generate the State table

count	A	B	C	D	A(t+1)	B(t+1)	C(t+1)	D(t+1)	TA	TB	TC	TD
0	0	0	0	0	0	0	0	1	0	0	0	1
1	0	0	0	1	0	0	1	0	0	0	1	0
2	0	0	1	0	0	0	1	1	0	0	0	1
3	0	0	1	1	0	1	0	0	0	1	1	1
4	0	1	0	0	0	1	0	1	0	0	0	1
5	0	1	0	1	0	1	1	0	0	0	1	1
6	0	1	1	0	0	1	1	1	0	0	0	1
7	0	1	1	1	1	0	0	0	1	1	1	1
8	1	0	0	0	1	0	0	1	0	0	0	1
9	1	0	0	1	0	0	0	0	1	0	0	1
10 to 15	Don't	Cares										

Step 3.b Generate the T Flip Flop Input Equations from the Table.

$TA = \sum m(7,9), d(10 \text{ to } 15)$

$TB = \sum m(3,7), d(10 \text{ to } 15)$

$TC = \sum m(1,3,5,7), d(10 \text{ to } 15)$

$TD = \sum m(0,2,3,4,5,6,7,8,9), d(10 \text{ to } 15)$

Step 3.b.1 Simplify using Karnaugh maps

The following are the simplified functions :

TA = B'CD + AD

TB = CD

TC = CD + A'D

TD = B + C + D'

Step 4 : Draw the Logic Diagram

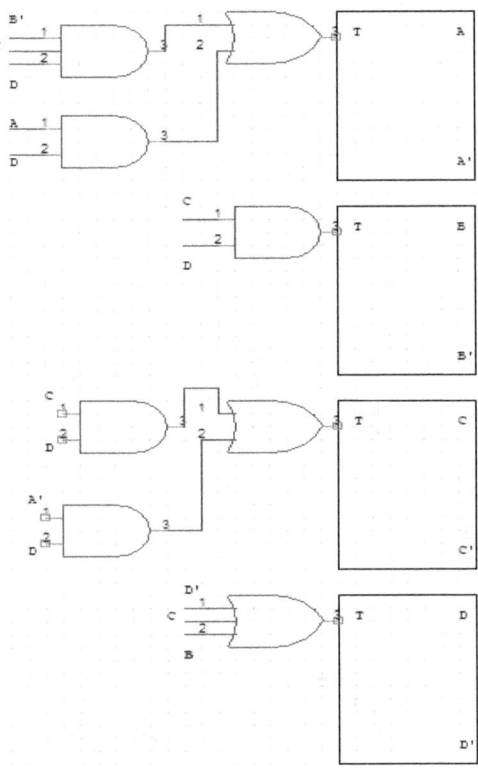

An OBE Approach to Logic Circuits and Digital Design: Step by Step

6 : Repeat IP#4 using JK Flip Flops

Step 2 : Generate the table.

count	A	B	C	D	A(t+1)	B(t+1)	C(t+1)	D(t+1)	JA	KA	JB	KB	JC	KC	JD	KD
0	0	0	0	0	0	0	0	1	0	X	0	X	0	X	1	X
1	0	0	0	1	0	0	1	0	0	X	1	X	1	X	X	1
2	0	0	1	0	0	0	1	1	0	X	1	X	X	0	1	X
3	0	0	1	1	0	1	0	0	0	X	1	X	X	1	X	1
4	0	1	0	0	0	1	0	1	0	X	X	0	0	X	1	X
5	0	1	0	1	0	1	1	0	0	X	X	0	1	X	0	X
6	0	1	1	0	0	1	1	1	0	X	X	0	X	0	1	X
7	0	1	1	1	1	0	0	0	1	X	X	1	X	1	X	1
8	1	0	0	0	1	0	0	1	X	0	0	X	0	X	1	X
9	1	0	0	1	0	0	0	0	X	1	0	X	0	X	X	1
10 to 15	Don't	Cares														

Step 3.b Generate the JK Flip Flop Input Equations in SOP abbreviated format:

$JA = \sum m(7), d(8 \text{ to } 15)$

$KA = \sum m(9), d(0 \text{ to } 7, 10 \text{ to } 15)$

$JB = \sum m(1,2,3), d(4 \text{ to } 7, 10 \text{ to } 15)$

$KB = \sum m(7), d(0 \text{ to } 3, 8 \text{ to } 15)$

$JC = \sum m(1,5), d(2,3,6,7,10 \text{ to } 15)$

$KC = \sum m(3,7), d(2,3,6,7,10 \text{ to } 15)$

$JD = \sum m(0,2,4,6,8), d(1,3,7,9, 10 \text{ to } 15)$

$KD = \sum m(1,3,7,9), d(0,2,4,5,6, 10 \text{ to } 15)$

Step3.b.1 Simplify the expressions using Karnaugh maps:

The simplified functions are :

$JA = BCD$

$KA = D$

$JB = C + A'D$

$KB = CD$

$JC = A'D$

$KC = D$

$JD = D'$

$KD = A'B' + CD + AD$

An OBE Approach to Logic Circuits and Digital Design: Step by Step

Step 4 : Draw the Logic Diagram

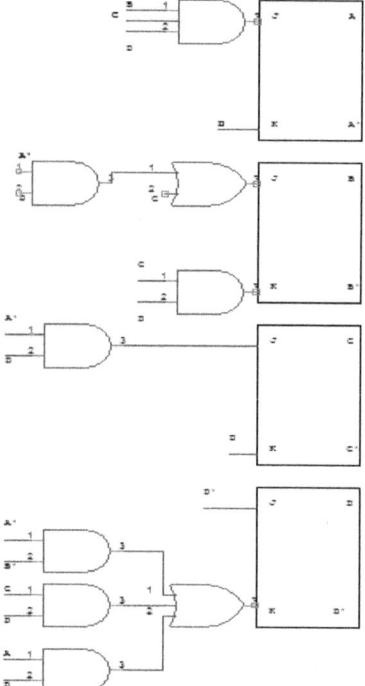

CHAPTER 11 END OF THE CHAPTER ACTIVITIES:

1. Design a Count Down Counter (9 to 0)

    a) Using D flip flops

    b) Using T flip flops

    c) Using JK flip flops

2. Design a shift left register.

3. Design a shift right register

4. Design a carry look ahead adder.

5. What is a Master Slave Flip Flop Design a SR Master Slave Flip Flop

CHAPTER 12 - NEXT STEPS : INNOVATE, INNOVATE, INNOVATE!

LEARNING OUTCOMES:

1. To demonstrate knowledge of the concepts illustrated in the previous chapters via a working model.

2. To analyze and design digital circuits whether it be combinational logic circuits or sequential logic circuits.

3. To apply in real life situations the concepts learned.

This chapter is yours to write. We walked through the last eleven chapters preparing you for this final activity. As an OBE book, the true test is in the outcome. The final goal is to give you the skill to illustrate what you have learned to solve real life problems and situations. To fill a gap or a need, that is the mother of all inventions and innovations. Supplement with research interfacing and I/O needs such as maybe the use of sensors or other form of displays aside from the seven segment display. Remember this is introductory and the intention is to instill in you the interest to pursue more complicated problems as you master the easier ones. Document your activities with pictures and videos. Make a storyboard or timeline of your progress.

Step 1 .STATEMENT OF THE PROBLEM (Determine right away whether it is a CLC or SLC problem)

_____

_____

Step 2 . Document the Steps taken to solve the problem . Write down all equations, tables, diagrams . take screenshots of your simulations and draw your circuit.

Attach additional pages . The more tools you use from the chapters, the better. Follow the steps as required.

Step 3 : Document your output with pictures. Paste pictures of your working model including screenshots of output simulation.

Step 4 : REFLECTION – How would you assess the entire book experience? Are you now more like Fleming or Kilby or Noyce or an entirely new breed of innovator? Share you insights. Use additional paper as necessary.

Notes

1. Dictionary . com , "digital electronics", in *The Free On-line Dictionary of Computing.* Source location Denis http://dictionary.reference.com. Accessed: September 19, 2014.

2. Dictionary . com, "logic gates," in *Dictionary .com Unabridged.* Source location : Random House, Inc. http://dictionary.reference.com/browse/logic gates. Available: http://dictionary.reference.com Howe. http:// dictionary.reference.com/browse/digital electronics. Available:. Accessed: September 19,2014.

3. Dictionary . com , "flip-flops," in *Dictionary .com Unabridged.* Source location : Random House, Inc. http://dictionary.reference.com/browse/flip-flops. Available: http://dictionary.reference.com. Accessed: September 19,2014.

4. Dictionary . com , "semiconductor," in *Dictionary .com Unabridged.* Source location : Random House, Inc. http://dictionary.reference.com/browse/semiconductor. Available: http://dictionary.reference.com. Accessed: September 19,2014.

5. Dictionary . com , "silicon" in *Dictionary .com Unabridged.* Source location : Random House, Inc. http://dictionary.reference.com/browse/silicon. Available: http://dictionary.reference.com. Accessed: September 19,2014.

6. Dictionary . com , "transistor" in *Dictionary .com Unabridged.* Source location : Random House, Inc. http://dictionary.reference.com/browse/transistor. Available: http://dictionary.reference.com. Accessed: September 19,2014.

7. Dictionary . com , "integrated circuit" in *Dictionary .com Unabridged.* Source location : Random House, Inc. http://dictionary.reference.com/browse/integrated circuit. Available: http://dictionary.reference.com. Accessed: September 19,2014.

8. Dictionary . com , "tipping point" in *Dictionary .com Unabridged.* Source location : Random House, Inc. http://dictionary.reference.com/browse/tipping point. Available: http://dictionary.reference.com. Accessed: September 19,2014.

9 "Fleming, John Ambrose." Encyclopedia of World Biography. 2005. encyclopedia.com (September 19, 2014) http://www.encyclopedia.com/doc/1G2-3446400070.html

10. "List of vacuum tube computers." Wikipedia (September 19, 2014) http://en.wikipedia.org/wiki/LIst_of_vacuum_tube_computers

11. "Transistor." ,Wikipedia (September 22, 2014) http://en.wikipedia.org/wiki/Transistor

12. "Integrated Circuit", www.nobelprize.org. (September 22, 2014) http://www.nobelprize.org/educational/physics/integrated_circuit /history

13. "Jack Kilby", Wikipedia (September 22, 2014) http://en.wikipedia.org/wiki/Jack Kilby

14. "Boolean Algebra", Wikipedia (November 11, 2014) http://en.wikipedia.org/wiki/Boolean_algebra

15. "Boolean Algebra", math.tutorvista.com(November 11,204) http://math.tutorvista.com/algebra/boolean-algebra.html

16. "Boolean algebra postulates and theorems", www.eecs.berkeley.edu (November 11, 2014) http://www.eecs.berkeley.edu/~newton/Classes/CS150sp98/lectures/week4_1/sld010.htm

17."What is the difference between a theorem and postulate", www.answers.com (November 11, 2014)http://www.answers.com/What __is_the_difference_between_a_theorem_and_postulates

References

DE GUZMAN, ESTEFANIA. *OBE Field Study 5 : Learning Assessment Strategies*, Quezon City: Adriana Publishing 2014

HAYES, J.P. *Introduction to Digital Logic Design,* Reading, MA : Addison Wesley, 1993

KIME, C. R. and MANO, M.M. *Logic and Computer Design Fundamentals*, Upper Saddle River, NJ: Prentice Hall 1997

MANO, M.M. *Digital Design*, 2nd ed. Englewood Cliffs, NJ: Prentice Hall 1991

ROTH, C.H. *Fundamentals of Logic Design,* 4th ed. St Paul : West, 1992.

TAUB, H ., and D. SCHILLING, *Digital Integrated Electronics,* New York; McGraw-Hill, 1977

INDEX

2- Variable K-maps 64
3-Variable K-maps 67
4-Variables K-map 71
abbreviated form 50
ADDERS..103
   half adder......................................130
ANALYSIS
   CLC..83
   slc 153
AND GATE...9
Bardeen 4
Boolean Algebra....................................23
Brattain...4
Characteristic table.............................146
Characteristic Table
   D flipflop.......................................147
   JK 157
   T 159
CLOCKED SR FLIP FLOP 142
CODE CONVERTERS...........................97
Complement of a Function....................25
Complements of Functions 23
Cost table..29
counter..183
D Flip Flop 146
DECODERS..109
DEMULTIPLEXER...............................123
DESIGN
   clc 96
   slc 176
Digital Electronics 2
Don't Cares 76
Duality..23
ENCODERS..114
excitation table 177
Exclusive – OR....................................11
*Flip Flop*2, 140, 142, 143, 144, 145, 146, 147, 149, 155, 157, 159, 176, 177, 180, 181, 182, 186, 189
FLIP FLOP TRIGGERING 148
FLIP-FLOPS..140
FULL ADDER...84
Full Subtractor.....................................107
FULL-SUBTRACTOR.........................86

history of digital electronics.................3, 7
Integrated Circuit.......2, 3, 6, 14, 111, 195
INTEGRATED CIRCUITS 5
INVERTER...9
KARNAUGH................................64, 80
KARNAUGH MAPS..........................64
Kilby...5, 6, 194
LEVELS OF INTEGRATION 14
Logic gate..2
logic gates..8
LSI...14
Maxterm 48
Maxterms...52
MCLC...35
MINIMIZATION 64
Minterm 48
minterms..49
Moore..6
MSI...14
multilevel Combinational Logic Circuit 35
MULTIPLEXERS.................................116
NAND GATE.....................................10
NOR GATE..10
Noyce...6
OR GATE..8
Postulate..23
Principle of Duality 24
PRIORITY ENCODER 115
PROCEDURE FOR CONVERSION BETWEEN FORMS 54
Product of Sums 48
PRODUCT OF SUMS (POS)................52
Product of Sums Expression 57
Product Term 48
SEQUENTIAL LOGIC CIRCUIT.......139
seven segment decoder........................100
Shockley..4
SIMPLIFICATION OF BOOLEAN EXPRESSIONS 26
Simplifying Product of Sums.................74
SLC 139
SSI...14
Standard Form......................................48
State Table...154

SUBTRACTORS 106	Transistors 4
SUM OF PRODUCTS 49	advantages 5
Sum of Products (SOP) 48	truth table 17
Sum of Products Expression 56	Truth Table 8
Sum Term 48	universal gates 8, 11, 35
THE JK FLIP FLOP 143	UNIVERSAL GATES 11
THE S-R (Set-Reset) Flip Flop 140	Vacuum Tube 3
THE T-FLIP Flop 146	Vacuum Tube Computers 4
theorem 23	**VLSI** 14

www.ingramcontent.com/pod-product-compliance
Lightning Source LLC
Chambersburg PA
CBHW080909170526
45158CB00008B/2053